Lecture Notes in Networks and Systems

Volume 113

Series Editor

Janusz Kacprzyk, Systems Research Institute, Polish Academy of Sciences,
Warsaw, Poland

Advisory Editors

Fernando Gomide, Department of Computer Engineering and Automation—DCA,
School of Electrical and Computer Engineering—FEEC, University of Campinas—
UNICAMP, São Paulo, Brazil
Okyay Kaynak, Department of Electrical and Electronic Engineering,
Bogazici University, Istanbul, Turkey
Derong Liu, Department of Electrical and Computer Engineering, University
of Illinois at Chicago, Chicago, USA; Institute of Automation, Chinese Academy
of Sciences, Beijing, China
Witold Pedrycz, Department of Electrical and Computer Engineering,
University of Alberta, Alberta, Canada; Systems Research Institute,
Polish Academy of Sciences, Warsaw, Poland
Marios M. Polycarpou, Department of Electrical and Computer Engineering,
KIOS Research Center for Intelligent Systems and Networks, University of Cyprus,
Nicosia, Cyprus
Imre J. Rudas, Óbuda University, Budapest, Hungary
Jun Wang, Department of Computer Science, City University of Hong Kong,
Kowloon, Hong Kong

The series "Lecture Notes in Networks and Systems" publishes the latest developments in Networks and Systems—quickly, informally and with high quality. Original research reported in proceedings and post-proceedings represents the core of LNNS.

Volumes published in LNNS embrace all aspects and subfields of, as well as new challenges in, Networks and Systems.

The series contains proceedings and edited volumes in systems and networks, spanning the areas of Cyber-Physical Systems, Autonomous Systems, Sensor Networks, Control Systems, Energy Systems, Automotive Systems, Biological Systems, Vehicular Networking and Connected Vehicles, Aerospace Systems, Automation, Manufacturing, Smart Grids, Nonlinear Systems, Power Systems, Robotics, Social Systems, Economic Systems and other. Of particular value to both the contributors and the readership are the short publication timeframe and the world-wide distribution and exposure which enable both a wide and rapid dissemination of research output.

The series covers the theory, applications, and perspectives on the state of the art and future developments relevant to systems and networks, decision making, control, complex processes and related areas, as embedded in the fields of interdisciplinary and applied sciences, engineering, computer science, physics, economics, social, and life sciences, as well as the paradigms and methodologies behind them.

** Indexing: The books of this series are submitted to ISI Proceedings, SCOPUS, Google Scholar and Springerlink **

More information about this series at http://www.springer.com/series/15179

Erik W. Aslaksen

The Stability of Society

 Springer

Erik W. Aslaksen
Gumbooya Pty Ltd
Allambie Heights, Sydney, NSW, Australia

ISSN 2367-3370   ISSN 2367-3389   (electronic)
Lecture Notes in Networks and Systems
ISBN 978-3-030-40225-9   ISBN 978-3-030-40226-6   (eBook)
https://doi.org/10.1007/978-3-030-40226-6

This Springer imprint is published by the registered company Springer Nature Switzerland AG
The registered company address is: Gewerbestrasse 11, 6330 Cham, Switzerland

Contents

Symbols

β	Level of stress
γ	Strength of the identity
ϑ	The set of assertions in Θ
Θ	Identity (a subset of memory)
κ	Commonality
μ	Information flow
ν	Strength of an assertion
σ	Predicate in an assertion
Q	The union of individual ϑs
r	Restraint
S	Subject in an assertion
S^σ	An assertion with subject S and predicate σ
s(ν)	Number of assertions with ν arguments
u	Mental processing capacity
w	Number of information items in Θ
$X(S^\sigma)$	Set of all arguments in Θ associated with S^σ
z	Number of assertions in Θ

Chapter 1
Introduction

Abstract The current drift in the international community towards polarisation is the motivation for this essay, which builds on a view of society presented in earlier work and on the idea of the evolution of society as the current stage in a general process of evolution. A number of assertions are presented, setting the stage for the subsequent development of a high-level model of society.

1.1 Motivation for This Essay

An earlier publication, *The Social Bond* [1], was concerned with how the interaction between the members of society drives its evolution. It presented society as an information-processing system with individuals as the distributed processors, and proposed a high-level model of the information processing and storage capabilities of the individual. The greater part of the book was then dedicated to studying the interaction between individuals; its measurement, its features, its dynamics, and its dependence on technology. Several simple models of the interaction were used as illustrations, and then, in the last chapter, some implications of this view of society's evolution for politics and economics were discussed. One of these implications—for the *stability* of the evolution—was discussed only very briefly, but in view of the rising tensions both within nations and on the international level, and a drift towards a confrontational behaviour of the currently most powerful nation, it seemed both appropriate and urgent to revisit the issue of the stability of the evolution of society. The present essay does that, and while it repeats some of the ideas and insights of *The Social Bond* for the sake of the reader's convenience in making it more self-contained, the focus is now firmly on the issue of stability, and all the supporting material—models and references—is presented with that in mind.

The issue of stability arises quite naturally if we consider the evolution of society to be the current phase of a general process of evolution and inheriting certain characteristics of that process, as proposed in the following section. The extent to which you find that view of evolution convincing will probably depend on your background, and the same is true of the various analogies between society and systems within physics and engineering used throughout this work as means of explanation

© The Editor(s) (if applicable) and The Author(s), under exclusive license to Springer
Nature Switzerland AG 2020
E. W. Aslaksen, *The Stability of Society*, Lecture Notes in Networks and Systems 113,
https://doi.org/10.1007/978-3-030-40226-6_1

and visualisation. Obviously, these analogies have to be at a high level of abstraction, screening out all lower levels of increasing detail, and each analogy is a compromise between explanatory power and complexity. It is less a question of their accuracy in any particular case, than of the range of cases in which they provide useful insight.

It will also become obvious that I am not a sociologist by profession, and that therefore the location of this essay in social theory, and the references and interfaces to the existing extensive body of work will appear deficient to sociologists. Again, I can only hope that it will be useful, and be seen as an invitation to critique and as an item for discussion and further development.

It is with gratitude that I acknowledge the support of the Faculty of Engineering and Information Technology at the University of Sydney in providing access to the University Library; the valuable comments and insights provided by Albrecht Fritzsche from the Friedrich-Alexander-Universität Erlangen-Nürnberg as part of his review of the draft manuscript; and, as always, the support of the other half of the team—my wife, Elfi.

1.2   A General Concept of Evolution

Viewing evolution as a particular type of transformation, as presented below, is based on considering *energy* as an abstract concept that appears in two forms: free energy and bound energy. Both forms are measured in terms of their intensity, or energy density, and because energy is conserved, an increase in one form of energy must result in a corresponding decrease in the other; it is always a conversion of one form into the other. The essential difference between the two forms of energy, no matter what their physical representations are in any particular case, is that the balance between free energy and bound energy is a function of the intensity of the free energy, with free energy dominating at high intensities. This behaviour is reflected in many circumstances that we are familiar with in daily life, where the free energy is in the form of heat and its characterisation is in terms of temperature, and for this reason we shall generalise the concept of temperature to be the intensity of the free energy in our abstract concept of energy. And, finally, it is important to realise that the conversion can go in both directions, so that bound energy created at one temperature will be converted back into free energy if the temperature rises, as, for example, in a fluctuation.

The purpose of introducing this abstraction of our well-known concept of energy is to assist us in forming a high-level view of evolution and, within evolution, of the history of society. So let us start at the beginning of our universe—the Big Bang—when a huge amount of energy with almost infinite intensity appeared, and it was all in the form of free energy. Being unconstrained, this energy propagated outwards, and as the temperature (i.e., energy density) decreased, the concept of *time* was created—time is a measure of change. But there was no concept of space, until the temperature decreased to the point where the conversion of free energy into bound energy became possible. The bound energy first appeared in the form of *particles* of various types—electrons, protons, and neutrons—and the concept of extension, and with it *space*, was created. As the temperature decreased further, these particles could *interact* to form increasingly complex entities of bound energy, in the form of *atoms*.

Specialise now to that part of the Universe that is of particular relevance to us—the planet Earth. At the point in the evolution where the Earth could be identified as a planet in the solar system, all the different atoms making up the matter of the Earth had been created; the development had come to a halt due to the instability of larger atoms. At first the planet was a sphere of hot gas, but as it cooled down, larger and larger combinations of atoms became possible; first as liquids, then also as solids, and the Earth was formed as a solid crust over a hot gas/liquid core, with gases being expelled through this crust from the ongoing compression of the core. Further cooling of the surface of this crust allowed water molecules to condense to water, and oxygen and nitrogen molecules to form the Earth's atmosphere. But, again, the formation of larger units of bound energy continued through the process of *interaction* of atoms, creating ever larger and more complex *molecules* as the temperature decreased.

As new molecules grew in size and complexity, and the decrease of the temperature slowed down more and more, a point was reached where further growth as individual molecules was no longer a viable option. Again, the way forward in converting free energy to bound energy was *interaction*, this time between molecules, but with one great change: The interaction was no longer in the form of a force between the participants but in the form of an exchange of matter. That is, what bound the molecules together was a *process* that enabled the entity to survive the inevitable fluctuations in temperature; a process that included an exchange of matter with the environment of the entity (metabolism). As these entities grew in size and complexity, the process of repairing the damage caused by the temperature fluctuations became an increasing overhead on the existence of the entity, and there came a point (well known to engineers) where replacement became more favourable than repair, so a process of replication was included as the entities evolved, and evolution arrived at what we call *cells*.

Various types of single-cell organisms developed (just as different atoms and molecules did earlier), but again, at some stage the increase in complexity came to an end, as the better way forward was to form new entities through interaction of different cell types, resulting in many of the organisms we see around us, such as plants. As these organisms increased in the complexity of their internal organisation and in the sophistication of their interaction with the environment—a complexity which demanded some form of central control, and an interaction which could now include sensing and movement—the susceptibility to fluctuations in temperature again became a problem, and the way forward was the development of a process and structure that could maintain a constant internal temperature, and this then allowed the evolution to progress all the way to its current state. However, in that last part of the evolution, there was a development that constitutes a separate and distinct step, and that was the development of the ability to communicate through speech—the particular characteristic of the genus *homo*, and a development of which we are the current, splendid result: *homo sapiens*.

What I want you to take forward from this brief account, which is greatly simplified and in which I have taken numerous liberties, is that evolution can be seen as having progressed through a number of *stages*, identical in that each one consists of two *phases*—a growth phase and an interaction phase—as summarised in Table 1.1. Throughout the greater part of this evolution, the generalised temperature decreased together with that of the environment, and the complexity of the emerging entities increased, through a conversion of free energy into bound energy, to the extent that they could survive the fluctuations of the environment's temperature. The growth of the complexity of the entities with the growth of their size can also be interpreted as the growth of the information contained in them, just as a binary number is both more complex and contains more information the longer it is (excluding any periodicity). The idea that evolution can be seen as an increase in information contained in the entities in each stage, and the role of this information in driving the evolution forward, are both strong motivators for the approach to evolution presented in this essay. At stage 4 in the evolution, the amount of information became sufficient to allow reproduction (as surmised by John von Neumann in the 1940s [2]) and thus reproduce the information that determines the hereditary properties of living things; we may now be at the dawn of a further such transition: into an era where we can not only reproduce the information, but also change it. The entities that emerged from evolution are now masters of evolution.

In the early stages the fluctuations in (generalised) temperature were overcome by the temperature difference between creation and destruction of the entities; in the later stages by homeostasis, fabrication (clothing, dwellings), and defence (in the case of societies). And where we have used the expression "the better way forward", the criterion for deciding what was "better" was always the same: the one most likely to *survive*. It is this high-level view of evolution that we want to apply in order to gain an understanding of its most recent phase—the evolution of societies through interaction of humans over the last 10,000 years or so.

Table 1.1 Evolution as a succession of stages, each consisting of two phases, with the transition from the first to the second resulting in the creation of a new type of entity trough interaction; i.e., through the conversion of free into bound energy

Stage	Entities evolving in the growth phase	Entities arising in the interaction phase	Nature of interaction
1	Free energy	Particles	Nuclear force
2	Particles	Atoms	Electromagnetic force
3	Atoms	Molecules	Covalent binding
4	Molecules	Cells	Exchange of matter
5	Cells	Organisms	Functional diversification
6	Organisms	Animals	Central control
7	Genus homo	Societies	Exchange of information

What is not shown, is the decrease in the duration of each stage

Table 1.2 The progression of societies and their constituent elements over time

Elements	Society
Individuals	Family
Families	Clan
Clans	Fiefdom
Fiefdoms	Kingdom
Kingdoms	Nation
Nations	World society

Within this framework, we can view the evolution of society in a similar manner, consisting of stages with two phases, and for this purpose it is most convenient to, for the moment, abstract from any particular society and simply consider a society as a system with its elements. In the first phase, as the system grows in size, there occurs a subdivision, or structuring, into entities that are distinguished by differences in the interactions between their elements. In the second phase, as the number of these entities grows, the interactions between them form them into a new system, on what we might consider a higher level and with the entities as the new elements. In the next stage, these two phases are repeated, resulting in new (larger, more complex) entities and a further (higher level) system. In terms of societies, this progression is shown in Table 1.2.

In Table 1.2, the transition from fiefdom to nation via kingdom was in many cases a conversion of a ruling monarchy into a constitutional monarchy; Germany is a case where kingdoms were merged into a nation. We should also note that as new elements are created, the elements of the previous stages do not disappear; they only assume new functions on another level of systemic organisation. If we, in our view of evolution, compare the development of single cells into multi-cellular organisms with the development of humans into societies, the increasing integration of the cells into the organism and their differentiation to form the increasingly complex structure of the organism is reflected in the integration of the individual into society and adaption to its increasingly complex structure. But just as the cell retained its importance as the building block of society, the same is true of the individual and of the sequence of structural components arising in the course of the evolution of society. In particular, both families and nation-states retain their identity and importance as they adapt to that evolution; the idea of a structure-less society is utopian.

The evolution we are concerned with is not its most common connotation—the evolution of species, as introduced by Charles Darwin and involving a process of largely random mutations, sexual reproduction, and a selection based on competition for resources and adaptation to changing environmental conditions—but if we consider the various societies that have existed, from small groups of hunter-gatherers to the present day nation-states, they may be considered as the species in an evolutionary process of the genus *society*, subjected to a similar process of mutations and competitive selection, except that the mutations were not random, but generated by the societies themselves. When we then consider what happened to the evolution of other parts of the animal world, e.g. the dinosaurs, and that most of the species that have existed are now extinct, we need to be concerned about the fate of humanity, now irreversibly embedded in the genus *society*.

However, while we may view the evolution of both the genus *homo* and of the genus *society* as phases in the above general process of evolution, there is a factor that is not present in such fields as, for example, physics and chemistry, and that is our consciousness; our view of ourselves as something special—and especially important—rather than just representing the current stage of an ongoing evolution. This is one reason for religion; a means of distancing us from the rest of evolution by postulating a special relationship to a divine sphere, but it is also reflected in other attitudes, such as political ideologies. It is reflected in our ambivalent attitude to society; we recognise it as being essential to our existence, but we resent it as a limitation on our divine right as individuals. The belief in a divine association in any form, and its unscrupulous exploitation by various organisations, has had a major influence on the evolution of society, manifested mainly in a suppression of knowledge and a delay in its application. It is not difficult to imagine a shaman exhorting his flock to abstain from making fire—it was clearly a prerogative of the gods, as could be seen when lightning struck—and it was only when a brave soul decided it was better to risk the wrath of the gods than to freeze to death that man began to use and control the power of fire.

And it is not all that different today. If we, in our view of evolution, recognise two discontinuities—the step from inert matter to living organisms, and the step from being subject to Nature to being the master of Nature—then it is pretty obvious that we are now looking at the third discontinuity, which will be going from accepting our own nature as given to being able to change it by genetic engineering, and that humanity will, at some point, not look anything like what it does today. But the divine-inspired resistance is alive and well.

1.3   Assertions

This essay attempts to present an approach to understanding our society that avoids any assumptions about the purpose or an end-state of evolution. However, as will become apparent as you progress through this essay, the approach is based on a number of assertions, and while their further elaboration and justification will emerge as our view of society is developed, they are stated here in order to serve as an anchor point for the various views of society that together will form our understanding.

> *First assertion*: Even at such a high level of abstraction as considering society to be a collection of interacting individuals, without any further characterisation or structuring of either the individuals or the interactions, we can determine useful and exact characteristics of society.

It is a common fallacy to equate high level with a lack of precision, and that as one progresses downward through levels of decreasing abstraction and increasing detail, the high-level characteristics will be shown to be approximations or simply incorrect. This arises because when high-level characteristics are observed, we tend to speculate on what the underlying causes might be, and it is these speculations that may prove to be partly, or even completely, incorrect as we develop our understanding in more detail.

Second assertion: All changes that take place in society, including those that lead to the evolution of society, are the result of human actions.

That is, while there may be natural (non-human induced) changes to the environment in which society exists, such as ice ages, how society reacts to such changes is determined by its members. Society is a product of human effort; its evolution reflects choices made by humans without any divine intervention. We made it what it is today, and we determine what it will be tomorrow. It is this view of society, as a process rather than a particular collection of objects, and of its members as the drivers of this process, that is fundamental to our approach to gaining an understanding of it.

Third assertion: The human actions can be divided into two groups: Those that sustain the current state of society, and those that change it. Because of their cyclic nature, these two groups will be identified as the *economic cycle* and the *transformation cycle*.

This assertion is perhaps the most significant and, potentially, the most controversial aspect of our approach to understanding society, as many people would say it is impossible to separate the two; any action will have some influence on the evolution of society (e.g. the butterfly effect). Nevertheless, this division will prove to be very useful; we only need to be careful to recognise its limitations.

Fourth assertion: Our actions are based on the *information* available to us at the time of deciding on the actions, where here and throughout this essay we shall understand information in the widest sense, as a concept that includes data, knowledge, understanding, belief, instinct, hope, wish, feeling, and the like.

Support for this assertion is the fact that people can be made to do anything, including sacrificing their own lives, if they have the appropriate information. As a consequence of this assertion, it is possible to see society as a giant distributed information-processing system; this is a particular *view* of society. The essay presents a number of different views, each one allowing us to gain further understanding of society as a whole and of its evolution.

1.4 Organisation of the Material

The next chapter introduces some basic concepts that will be referred to in our description of society throughout this essay. In particular, the concept of a system, and its application in constructing different views of society, is central to the approach used to form high-level descriptions of such a complex entity as a society. Chapter 3 develops the principal view of society as an information-processing system, with a simple model of individuals as the system elements, and with the information flow between individuals determining the emergent behaviour, or evolution, of society. As what might be considered somewhat of a digression, it is shown how the features of our system model is reflected in some recent work on the development of settlements.

Chapter 4 then looks in more detail at the collective intelligence that drives this evolution, and how it forms the criteria of its decision-making process. Of importance is the assertion that these criteria themselves evolve; there is no end state towards which society should progress; our focus should be on the process, not on any perceived ideal form of society. It is the process that has an ideal form, and deviations from this ideal can be observed as fluctuations in certain parameters characterising the state of society.

The next three chapters examine the main features of society that influence its evolution—technology and education, as well as two processes characterising its state—economics and politics. These contain the activities that influence and sustain the information-processing, and in this manner determine the dynamics of the evolution. Society is not only a complex system but a dynamic one, and one whose rate of change is rapidly increasing, which leads naturally to a concern for its stability—the subject of the last chapter. There stability is considered on two levels—the nation-state and the global level—while pointing out that these are becoming more intertwined due to a combination of globalisation and what is known as "proxy wars". Having escaped from one high-risk situation in the form of the Cold War, we can now clearly see an underlying feature of the global structure that could be driving us towards a major fluctuation.

References

1. Aslaksen EW (2018) The social bond: how the interaction between individuals drives the evolution of society. Springer, Berlin
2. von Neumann J (1948) The general and logical theory of automata. Presented at the Hixon Symposium, and available online at http://physics.bu.edu/~pankajm/PY571/HixonsymposiumVonNeumaan.pdf

Chapter 2
Some Basic Concepts

Abstract Introduces the concept of a system as a mode of description, and applies it to a view of society consisting of individuals, their interactions, and their belongings (artefacts and knowledge). Discusses various approaches to how to provide a description of the state of society and its evolution, including a brief consideration of the activities performed by the members of society.

2.1 Systems and Views

A society, such as a nation-state or even the whole world, is a very complex entity, and different professional disciplines, including sociology, anthropology, political science, philosophy, and economics, are engaged in describing and developing an increasingly detailed understanding of it, documented in a large body of published work. As with all complex entities, it is easy to become focused on a description in terms of the concepts and methodologies of a particular discipline, gradually giving it an independent existence, thereby losing sight of the fact that its contribution and relevance to the entity as a whole arises only through its interaction with all other aspects. To avoid that, and to get a handle on this complexity and make a start on understanding what a society is, we can use a well-known approach, which consists of viewing a complex entity as a system, and then employ a number of existing tools or methodologies for analysing systems. This system methodology was originally developed as an aid in creating complex technological entities, such as telephone networks and missile defence systems, where it goes under the name of systems engineering, but it has since been generalised to apply to both the creation and analysis of any complex entity, and can simply be viewed as a means of handling complexity of any sort.

Formally, the *system concept* is defined in terms of three sets:

- a set of elements;
- a set of interactions between these elements; and
- a set of interactions with the outside world (which may simply be an observer).

The system concept is a *mode of description*; nothing *is* a system, but everything can be described as a system. For any particular object, we can choose what we

E. W. Aslaksen, *The Stability of Society*, Lecture Notes in Networks and Systems 113,
https://doi.org/10.1007/978-3-030-40226-6_2

identify as elements, and there are normally several possible descriptions of an entity as a system, depending on what aspect of the entity we are interested in examining; each such description is called a *view*. In this manner we are hiding the complexity that is not relevant to the aspect of interest, but because all the views are connected, in that they relate to the same entity, they themselves form a system. (For a detailed description, see [1]).

As a demonstration of this, consider a well-known household item: the refrigerator. From the point of view of its physical composition, we could describe the refrigerator as a system consisting of the following elements (units):

- a cabinet, with door, drawers, and shelves;
- a compressor unit, which compresses the gas and thereby heats it;
- an external heat exchanger (condenser), in which the gas is cooled by giving off heat to the ambient air, to the point where the gas liquefies;
- an internal heat exchanger, into which the liquid flash evaporates into cold gas and through which it absorbs heat from the air in the cabinet; and
- a control unit.

This is the physical view, sometimes also called component view or block view. Each of these units is itself quite complex, and requires numerous parameters in order to describe everything about it. But if we are interested in only one aspect of the refrigerator, we may ignore, or hide, all of these parameters except the one we are interested in. This might be the weight of the refrigerator, in which case we only need the value of this one parameter for each unit, or it might be the reliability, in which case we only need the failure rate of each unit, or it might be the cost, in which case we only need the cost of each unit. In each of these cases, the system structure is very simple; it is simply a linear chain with the interaction being addition.

However, if our purpose of describing the refrigerator as a system is to explain the process, we might delete the first and the last of the elements above, representing the refrigerator as a system of only three elements. The structure would now be a closed loop, and each element, as well as the interaction between them, would be described by a few thermodynamic variables and a number of equations relating them to each other. However, if we are interested only in the energy balance, we might chose to describe the refrigerator as a system of three very different (abstract, not physical) elements:

- an element representing all the energy input (electricity);
- an element representing the energy loss through the thermal insulation; and
- an element representing the energy loss (inefficiency) of the process;

each characterised by a single variable, energy. So, what this little example demonstrates is that for any entity or object (in this case the refrigerator) there are many different descriptions of it as a system, each one addressing a particular aspect. These systems may differ in the choice of elements, in the structure (i.e. the interactions between the elements), and in the parameters describing the elements, but they all relate to the same entity.

2.2 Description of Society

The word "society" is used to describe a wide range of associations between the people in an identifiable group. Some of them very loose and playing only a minor part in the lives of the people involved, others describing a strong interaction that constitute a significant aspect of the existence of the members. Among the former are such societies as most professional societies, special interest or hobby societies, and sporting associations, among the latter are religious societies, such as the Society of Jesus or the Society of Friends. All of these are characterised by the fact that the members belong because they want to belong; membership is an expression of the individual's will. The societies we are interested in, which are generally nations or, in some cases, the world's population as a single society, are of a different type. In these societies, people are members by virtue of being resident within the geographical area that defines the society; the extent to which an individual participates in the activities of the society varies considerably, and in some cases there is no definite requirement to participate in any of them at all. With a few explicit exceptions, in this essay we shall take "society" to be understood either as "nation" or as "world". Which one of these is applicable will be easily understood from the context.

Our initial description of society as a system is in terms of its physical components—what it consists of (neglecting external interactions)—and we define *society* as consisting of three components:

- the individual members—the *individuals*;
- their public actions—the *interactions* between them; and
- their *belongings*.

The belongings include artefacts, such as tools and utensils, equipment and devices, clothing, artworks, means of transportation, and dwellings; domesticated animals; physical infrastructure, such as arable land, waterways, railways, roads and bridges; and, most importantly, knowledge, contained in verbal traditions and in written and pictorial representations. They are identified as belongings because they belong to a particular society; they form part of the characterisation of that society.

The public actions of the individual members are those actions that establish some form of relationship with at least one other member. Obvious examples are those actions performed in public; a more subtle example would be someone sitting alone at home reading a book—the relationship is with the author. The opposite are private actions, such as sleeping or brushing teeth; they form part of the description of the individual. Public actions have a physical content, as in writing a letter or shaking hands, and an information content, as in the information contained in the letter or the acknowledgement expressed in the handshake.

The concept of an *individual*, as used in this essay, has a dual definition, much as the dual description of a photon as a particle or as a wave in quantum mechanics. On the one hand we define an individual as an average member of society; that is, as a human whose characteristics, such as intelligence, physical capabilities, and life cycle, take on values that are the average of these characteristics over all members of society (although we may, later on in this essay, limit this to members over a certain age). This is similar to the use of *per capita* parameter values in economics, and to this extent, the individuals are all identical. But the information contained within them is different for each one, and we are able to speak about the value of a characteristic of this information of a particular individual. For example, as we shall see later, the interaction between two individuals will depend on how much they have in common, which implies that there are things they do not have in common, so that, to this extent, they are not identical. But when we then take the average over all pairs, what two individuals have in common is independent of which two individuals are involved in the interaction. We might express this by saying that behind the individual sits a person who is obscured because of the averaging process.

The actions performed by the members of society are based on their physical circumstances and on the information available to them at the time. Leaving the physical circumstances aside for the moment, the information available to a member of society is a combination of information stored from previous experience and information received at the time, and the processing of the received information depends on the stored information and the member's processing power. The latter may vary somewhat from person to person, just as other physical characteristics do, but within the degree of accuracy of our high-level approach, we shall assume that the processing power is the same for all persons. It then follows that the result of the processing depends on the stored information only; that is, when two persons arrive at different interpretations of the same received information and take different adaptive actions, it is due to differences in their stored information, and not to any inherent differences in processing power.

Now, humans, viewed as organisms, have not changed significantly over the last 10,000 years or so; we look the same, our physical capabilities are much the same, and so is the processing power of the brain. Combining this insight with the foregoing, it follows that the behaviour of society is completely determined by the information contained within society, and so our approach to understanding society and, in particular, its evolution, will be based on viewing society as a giant information-processing system, with brains as the individual processors, which was already stated in Sect. 1.1 as a consequence of our fourth assertion.

This, then, in effect, becomes a further view of society as a system. It has the same three main components, but they are now defined by how they relate to information; it is no longer a physical view, but a *functional* view. The individual is an abstract component that processes information; the interactions are interchanges of information, and the belongings are characterised by how they influence the information exchange. This will be our primary view of society; its purpose is to provide a framework in which various aspects of society can be considered in greater detail in the following chapters without losing sight of their interrelationships.

2.3 Measures of Society

The state of society can be measured in terms of many different parameters, depending on what one intends to achieve by the measurement. One measure might be the size of society, measured by national or total world population; another measure might be the Gross National or World Product; a further one could be the number of books published each year, or the level of education world-wide, and so on. The extent of technology could in principle be an obvious measure, even though the definition of this measure is difficult due to the fuzzy meaning of "technology" in daily language. It could also be a combination of measures, but in any case, in addition to defining the state of society, the measures need to be able to express the direction of evolution, its dynamics, as that is the focus of this essay.

The operative word here is "direction". Consider, as a simple, but concrete example, the movement of a car. As we have a measure of distance—the metre, and a measure of time—the second, we can measure how far it moved in a given time, and, by dividing distance by time, we can determine the speed of the movement. We can also measure the direction in which it has moved, by comparing with a preferred direction, such as the direction of the road or a compass direction, and the measure is in angular degrees. For none of these measurements is it necessary to know where the journey will eventually end, or if it even has an end. It is a bit like Forrest Gump; the essence is in the journey, not in the goal. If we look back at society, evolution is very evident; it has taken us from the cave to where we are today. But is there a preferred direction, and if so, is this direction defined by a goal? Or is it just a random process with no discernible overall direction? Looking at human history over the last ten thousand years or so, there can be little doubt that the "richness" of life, as measured by the diversity of our concepts and experiences, both on the average and in total, has increased exponentially in historic times, despite many setbacks in the form of wars and ideological subjugation. In most of the world, the opportunities for self-fulfilment available to the average person through material well-being and the associated free (or non-working) time, education, and an intellectually stimulating environment have led to societies that are again promoting those same factors, while attending to the various issues, moral and otherwise, accompanying that development. While it may appear that "the good life" has become synonymous with material success; it must not be overlooked that the access to and participation in all forms of art and political, scientific, and religious discourse, as well as a greater understanding of our environment through education and widespread dissemination of information have greatly increased the non-material content of the average person's life. And although this does not tell us where we will ultimately end up, or whether this is even a question we can expect to get an answer to, we can determine the direction in which we have been heading; a sort of local or differential determination of direction. One question would then be: Is this a "good" direction? Most attempts to provide an answer to this question presuppose that it is to be sought in something external to us; by recourse to a divine or supernatural authority. But if we understand that we are now the masters of evolution, the answer must be sought in an understanding of the characteristics of that mastery—of the process that drives evolution.

A first perspective on our search for system variables arises from the concept of a generalised energy introduced earlier, and by associating this energy with the activities performed by the individuals. We can then define the bound energy as the energy associated with those activities that require cooperation between individuals and are determined by society's requirements, rules, structure, and other constraints on such cooperation; and free energy as the energy associated with those activities that do not depend on the cooperation of other individuals and are at the individual's discretion. For each individual we can form the ratio of bound to total energy, and average this over all individuals; the resulting quantity is one aspect of a parameter we may think of as the degree of *social integration* (the choice of this name will become clearer further on). If we then consider three states of society, as measured by this parameter, and form an analogy with three states of a physical system—specifically, a collection of water molecules—we can obtain the following picture:

Society		Water molecules	
Social integration	State	State	Temperature
Low	Individual hunter/gatherers	Gas (steam)	High >100 °C
Medium	Nomadic societies	Liquid (water)	Medium 0–100 °C
High	Settled societies	Solid (ice)	Low <0 °C

The immediate idea you might take away from this picture is that the inverse of the degree of social integration is what we can think of as a *system temperature*. This temperature of societies has been declining from a very high value towards a value of 1, and we might reasonably ask if there is an ideal value of the temperature of a society, and what would determine such a value. For a given society, too high a temperature would tend to cause disintegration; too low a temperature would lead to stagnation; in this sense a society is like a living organism. A society is a dynamic entity, constantly evolving, but there is no goal to this evolution, no target society; it is impossible to know what society will look like in a thousand years from now. What is important is not its form at any one time, but the process of change, and the temperature is an indication of the stability of this process.

Based on our view of society as a collection of interacting individuals, the possible measures fall into two groups: those that relate to the individual, which we might call *micro measures*, and those relating to the collection as a whole, i.e., to the environment which supports the existence of the individuals and provides the infrastructure that supports the interaction between them, which we might call *macro measures*, of which the temperature is one. To illustrate the difference between these two groups of measures, consider again the analogy with a collection of water molecules. The properties of the collection will depend on the properties of the individual molecules, such as dipole moment, but they will also depend on the state of the collection as a whole, with its three distinct phases of gas (steam), liquid (water), and solid (ice). The most common measure of this collective state is temperature; another measure is what we identified as free energy, with the conversion of free energy into bound energy (and vice versa) taking place at the phase transitions. Another measure would be the *freedom* of the individual molecule within the "society"; it ranges from completely free in the gas to completely bound in the solid, but it is obvious that this measure can be viewed as either a macro measure or a micro measure, and so we are led to the realisation that micro measures are further divided into two subgroups: *intrinsic* measures that characterise the individual as such, independently of any relations to other individuals, and *contingent* measures, that depend on the fact that the individual is embedded in a society. Or, in other words, there is a conceptual difference between an individual considered in isolation and an individual embedded in society, where it is interacting with other individuals; this is a reflection of the dual nature of the concept of an individual, as discussed in the previous section.

An important micro measure of the second kind, and one that we will consider from a number of perspectives, is related to *the individual's perception of its purpose*: the advancement of its own interests, or the advancement of the interests of society. The extreme is a person that perceives its purpose to be the advancement of its own interests only, without any other consideration. This is a person that sees society simply as a collection of individuals; as an environment with which it interacts for its own benefit only, much as we used to see Nature (and some people appear to still do). Such a society would be simply the sum of its members, without any emergent properties. Only in the most primitive of societies would this be a reasonable view; in a modern society it must be considered delusional, and the behaviour resulting from it would be completely asocial. In thinking about how we could assign a measure to this dimension, say, as the ratio of social interests to individual interests, we first need to realise that as society evolves in the direction of stronger interactions and greater complexity, not only do society's requirements on the individual's actions increase, but so do the options open to the individual, so that this measure is not the same as what we think of as progress in the form of increasing GDP *per capita*. It is possible (likely?) that in our society today, the options offered to the individual are increasing faster than society is able to introduce requirements, and that the evolution of society, along this dimension, is actually regressing.

The difficulty with, but also the importance of, this measure is that it is defined in terms of the individual's perception, something that is not immediately measurable or directly observable. Any determination of this measure, which is a new perspective on the degree of social integration introduced above, would have to be determined by taking a test, similar to an IQ test, and the design of such a test would be a project within social science. It is useful to compare this concept of social integration with the well-established concept of *socialisation*. The Internet offers a number of definitions (some slightly edited):

Socialization is the process of internalizing the norms and ideologies of society. Socialization encompasses both learning and teaching and is thus "the means by

which social and cultural continuity are attained". https://en.wikipedia.org/wiki/
Socialization.

Socialization is a continuing process whereby an individual acquires a personal identity and learns the norms, values, behavior, and social skills appropriate to his or her social position. http://www.dictionary.com/browse/socialization.

Process by which individuals acquire the knowledge, language, social skills, and value to conform to the norms and roles required for integration into a group or community. It is a combination of both self-imposed (because the individual wants to conform) and externally-imposed rules, and the expectations of the others. http://www.businessdictionary.com/definition/socialization.html.

Socialization, the process whereby an individual learns to adjust to a group (or society) and behave in a manner approved by the group (or society), and is a central influence on the behaviour, beliefs, and actions of adults as well as of children. https://www.britannica.com/topic/socialization.

Socialization is the process that prepares humans to function in social life. It should be re-iterated here that socialization is culturally relative—people in different cultures and people that occupy different racial, classed, gendered, sexual, and religious social locations are socialized differently. https://en.wikibooks.org/wiki/Introduction_to_Sociology/Socialization.

What all of these definitions have in common is that the degree to which a person is socialised is judged by society; it is a matter of the person adapting his or her *behaviour* to the expectations of society. Whereas the degree of social integration is a characteristic of the persons *belief* or world view, a private view that does not necessarily have to be revealed to society. In most cases the two measures will be close, but it is entirely possible for a person, who society judges to be highly socialised, to have a very low opinion of society, seeing it only as an environment to be exploited for personal benefit, and not as something of which he or she is an integral part.

Social integration is not the same as morality; a person with a low degree of social integration can have high moral standards and behave accordingly; the following example illustrates this. Consider persons with high moral standards and a strong work ethic, who are, and have been, working hard and are earning a great deal of money. They employ aggressive tax minimisation schemes, to the maximum extent allowed within the law, but they are compassionate about people in need and donate a significant portion of their wealth to charity, as well as to other causes they deem worthy, such as art, research, education, and sport. They see wealth redistribution by the government as usurping an important personal moral activity and duty, besides, perhaps, having little faith in the government's ability to operate cost-effectively. Disregarding any ulterior motives, such as the desire for recognition and influence that tends to accrue from this behaviour, many of these persons would be convinced that they are acting in the best interests of society.

The issue here is not one of morality or of selfishness; it is one of understanding. The most immediate understanding, as individuals, of our relationship to society is as a relationship between individuals; between me and other individuals like me, with no reference to society. And so, if each one of us, individually, does what is right, society will be right. This is the personal morality that has been delivered to us from antiquity onwards, and as long as the intensity of the interactions between individuals was relatively low, this was a workable approximation, resulting in a zeroth-order view of society as the aggregate of individual contributions. But as the intensity of the interactions increased, societies displayed behaviours that could not be explained in terms of characteristics of individuals; the interactions introduced emergent features of the behaviour. It is then no longer enough to have a code of personal behaviour; we need a code of behaviour for society, and as the only active elements of society are individuals, it follows that each individual has, in addition to a responsibility for its own behaviour, a responsibility for the behaviour of society. The prerequisite for the individual taking on such a responsibility is that it has an adequate understanding of how the behaviour of society arises out of the behaviour of individuals and their interactions, just as we understand how the macroscopic properties of a substance arise out of the properties of its atoms (or molecules) and their interactions. This is not a simple relationship, nor is it one that is intuitive, given our origin as individuals; it is an understanding we have started to develop through experience and analysis over thousands of years.

What about macro measures of the evolution of society? We mentioned temperature earlier, although that is more a qualitative measure than a quantitative measure, and we introduced the concept of social integration. And in the next chapter we shall develop the concept of social integration further. But in the context of our view of society as an information-processing system, the significant macro measures will be measures of how the other activities taking place in society interact with and influence the information processing; these measures are already well defined and understood, and data for most of them is readily available. The next section presents a very brief overview of the major activities and their development in recent times, and in Chaps. 5, 6, and 7 we investigate their influence and introduce a number of global measures of these influences.

2.4   Activities

As another view, let us look at in what activities people spend their time, and to that end we might decide to consider all activities taking place in society to be grouped into five groups, as shown in Fig. 2.1. The interaction between these groups is that the sum of the time spent in each one of them must equal the total available time.

To illustrate the meaning of Fig. 2.1 by an example, let us look at a generalised, and somewhat hypothetical example of a developed nation, but one which is patterned on a real nation, Sweden, for which good statistical data is available (Swedish Bureau of Statistics, at www.scb.se/en/finding-statistics/). There, the lifetime of an average person is about 80 years. However, it takes some time after birth before the person accounts for activities and expenditure in its own right, so let us simply discount the first 5 years from this activity view. Of the remaining 75 years, a certain amount is

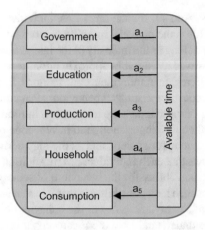

Fig. 2.1 The *activity view* of society; with the five activity groups in which the members of society (i.e. the population) spend their time. The definitions of the groups are as follows: Government: Legislative, executive, legal and law enforcement institutions, defence, customs, taxation, and other administrative services. Education: Teaching and studying at all levels. Production: Provision of goods and services, incl. research, development, manufacturing, sales, marketing, construction, operation, and maintenance, and incl. government owned operations (transportation, health, water, power, etc.). Household: All the household work that could be performed by paid staff. Consumption: All other activities. The five parameters, a_i, are the proportions of available time spent in each of the five groups over the lifetime of the average person

spent sleeping; say, on the average 8 h per day, so that the time available for spending on the five activities identified in Fig. 2.1 is 438,000 h.

Let the working portion of the lifetime be 45 years (from 20 to 65 years of age). From the Swedish data, the time spent working per person in this age range per week is 26 h (to which we add 6 h for commuting) or, over the lifetime, 74,880 h (this excludes household work), and this is distributed over the three activity groups labelled Government, Education, and Production in Fig. 2.1.

Each person might spend an average of 30 h per week, or 1200 h per year, over 15 years (from age 5 to age 20) on education, equating to 18,000 h. If 1 in 13 is engaged in teaching, this equates to another 4680 h on education, or a total of 22,680 h, so $a2 = 22{,}680/438{,}000 = 0.0518$. And if 1 in 5 is engaged in government activities (excluding education), $a1 = 12{,}168/438{,}000 = 0.0278$. As a result, it follows that the time spent on production is $74{,}880 - 4680 - 12{,}168 = 58{,}032$ h, and $a3 = 0.1325$. In this simple example we make no gender distinction, so it might be reasonable to say that each person spends 2 h per day between the ages of 25 and 75 on household work, or a total of 36,500 h, so $a4 = 0.0841$. We then have that the time remaining for consumption equals $438{,}000 - 74{,}880 - 18{,}000 - 36{,}500 = 308{,}620$ h, so $a5 = 0.7046$.

If we now want to consider these activities in the context of evolution, we need to take account of the decrease in life time going back in time, at least back to when some data is available, which for Sweden is about 1750. The life expectancy, or expected age at death, for persons at age 5 is shown in Fig. 2.2.

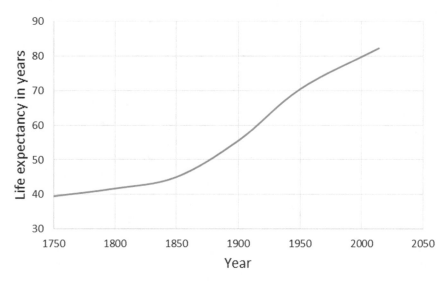

Fig. 2.2 Expected age at death for persons at age 5. This is an approximation due to the uncertainty regarding the extent of infant (i.e. <5 years of age) death rates in the early years

If we now look at the year 1750, the hours available to the average person (e.g. per capita) for the five activities in Fig. 2.1 were 204,400. The working portion of the lifetime would have been approximately 27 years (from 13 to 40 years of age), and with a six day working week at something like 10 h per day, the time spent working over the lifetime was 4680 h. Each person might have spent an average of 15 h per week for 6 years, equating to 4320 h, and teaching would have accounted for about 800 h (i.e. about same ratio of teachers/pupils as in 2000). It is difficult to determine the amount of time spent on household chores, as these were very much integrated into the work (e.g. as in agrarian production for own consumption), but in the absence of anything more definite, we might assume it to have remained basically constant. That is, the gain effected by various household equipment was absorbed by larger houses and higher standards. So, 2 h per day between ages 13 and 40, or 19,656 h. As a result, the time remaining for consumption was about 89,800 h.

From the above (very rough) calculations related to our activity view we only need to keep two results in mind as we progress: Over the period of 250 years, the time available for consumption, i.e. discretionary or "free" time increased by a factor of more than 3, and the time spent on education increased by a factor of more than 4.

Reference

1. Aslaksen EW (2013) The system concept and its application to engineering. Springer, Berlin

Chapter 3
The Information-Processing System

Abstract Specialises the view of society to that of an information-processing system. Central to this view is dividing the information into three classes, and the first consequence of this classification is the description of the related activities as forming two cycles, an economic cycle (daily life) and a transformation cycle (formation of belief). The information-processing individual is described in terms of a model, which introduces the central concept of identity and the associated structure of the process, and the interaction is extended to include the mediation by information technology. The relevance of work by R. Fletcher on the growth of settlements is discussed.

3.1 The System Components

In the view of society as an information-processing system, the definition of the three sets constituting the system are defined as follows:

- the *individuals* as information processors;
- the *interactions* as exchanges of information; and
- the *belongings* as the material environment in which the information processing and transmission takes place.

This definition reflects the fact that this view is concerned with *information*—its processing and transmission. The information represents society—its current state and its dynamics; and so we should start the development of this view by stating what our understanding of "information" will be.

The first step is to recognise that the information received by a person spans a wide range of subject matters, of importance, of truthfulness, and of relevance to the person, and that we will need to find some way of handling this diversity, not least because the total volume of information impinging on each person in a modern society is both very substantial and also increasing, as a result of developments in information technology. As a start, we subdivide the information into the following three *classes*:

E. W. Aslaksen, *The Stability of Society*, Lecture Notes in Networks and Systems 113, https://doi.org/10.1007/978-3-030-40226-6_3

1. The inputs that we either are not consciously aware of or do not actively engage with; perceptions of our environment when we are not focusing on anything specific; they are the inputs we associate with being awake. While some of these inputs are stored, and we might recall them later if prompted to do so, they do not result in any further processing, but they do take up some of the person's processing capacity, in the form of distractions.
2. The factual and uncontroversial inputs of everyday life, mostly connected with family life, work, education, and entertainment; inputs that, when evaluated against the contents of memory, result in the updating of memory without creating any conflicts or inconsistencies, and in predictable actions.
3. Inputs that relate to "the things that matter" [1] or, with reference to [2], "culture", which includes beliefs, attitudes, behaviour, art, and social norms. The evaluation of these inputs may result in a more or less direct and immediate adaptive (i.e., behaviour-changing) action, such as participating in a protest march, avoiding eating meat, writing a letter to the newspaper, etc., but may also raise conflicts with existing information items that then need to be resolved.

At this point it is appropriate to make a brief comment about an implication of this subdivision that arises out of the fourth assertion. According to that assertion, the subdivision of the information must be reflected in a subdivision of the activities taking place in society. Such a subdivision was introduced through the third assertion, and we can see that the *economic cycle* uses the information in class 2 to develop new applications of technology and associated products, and in the process generates new information and technology, while the *transformation cycle* uses the information in class 3 to generate new beliefs and corresponding adaptive actions, with this change to society resulting in new information in class 3. We asserted that the transformation cycle is the one controlling the evolution of society, with the economic cycle providing the environment in which it takes place, and thereby closing the feed-back loop between the two cycles, and we noted that this assertion is perhaps the most significant and, potentially, the most controversial aspect of our approach to understanding society. What it says is that the majority of our factual knowledge, such as the value of π, the number of planets of the Sun, names of rivers and cities, how to drive a car, and how to make steel, while very important for sustaining society, does not determine the direction of change of society. As an analogy, if one person shoots another person, the gun, as the item of technology, and thus the class 2 information, involved, is an enabler of the action, but the action is determined by the person's processing of the relevant available information (which would be mainly class 3). One could take the view that if the person did not have a gun, the event would not have taken place, and that therefore the gun, i.e., the technology, is the determining factor. That is not our view, nor is it that of our legal system; it is the person that is punished, not the gun. That is not to say, of course, that one should not be concerned about the availability of the gun, as is true of many applications of technology, but, again, any restrictions or procedures are the result of human decisions, i.e., of information processing, and mainly the processing of information in class 3.

A different perspective on the distinction between these three classes of the information flow is provided by the concept of *attention*—focused mental engagement with a particular information item. Class 1 does not require attention, whereas classes 2 and 3 do. Because of this, the concept of attention becomes a key factor in the discussion. And it will sometimes also be useful to recognise that the information can be classified along a different "dimension", according to whether it pertains to a process, i.e. to how to do something, such as extracting the square root of a number (which we might call an algorithm), or to a thing, such as "the Thirty Years War started in 1618". Under the perspective of education, the former requires practice, whereas the latter does not. There is a large body of work (in the cognitive sciences) relating to the different types of knowledge and to such aspects as attention, awareness, and conscious and unconscious behaviour. A small part of this work was considered in *The Social Bond*, but none of this detail is really necessary for the high-level view put forward in this essay.

Irrespective of the class of information, we introduce a common quantitative measure through the concept of an *information item*. Information items are the kind of statements and messages we would use in conversation with other persons, such as "I see we are expecting rain tomorrow", "the value of π is 3.1416", "I believe the world was created five thousand years ago", "you should not smoke", or an instruction about how to do something; and the fact that they vary greatly in complexity, potential impact, truth value, etc. is not relevant to our use of the concept. It is an average or a fiction, just as the average person is a fiction, and we will not need any knowledge of its nature, as we will only use it as a relative measure of quantity. In the cognitive and social sciences, information items are often referred to as *cognitions* or *mental states*. The term "information item" was chosen here for two reasons: One, in order to disassociate it from any further level of detail that may be attached to the use of the other two terms, and two, in order to allude to the association of our model with information technology and the brain as a computer.

3.2 The Individual

The information-processing capability of the individual is initially modelled in terms of the following simple model with only two elements: a processor and a memory, as shown in Fig. 3.1.

In this model, the Processor *evaluates* the incoming information on the basis of the knowledge stored in the Memory, and as a result takes what we shall call "adaptive action". This action might be an actual physical action, such as pressing a button, writing a message, or moving to another place; in our view of society as an information-processing system we are primarily interested in the information content of the action. This is the information output indicated by the arrow from Processor to Society in Fig. 3.1, but the action can also result in a change to the knowledge stored in the Memory, as we shall look at in greater detail in the next chapter, and which is the reason the arrow between Processor and Memory points in both directions.

Fig. 3.1 Two-component
model of the individual and
its two information
inputs—interaction with
society and discovery of the
environment

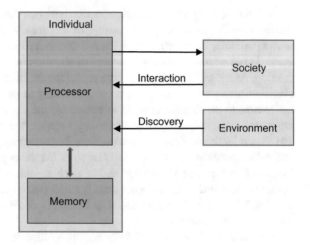

But already here we can see the difference to the more traditional applications of the
system concept, such as in engineering, where the system elements are completely
defined as stand-alone elements in terms of their characteristics, including their ability
to interact with other elements (i.e., their interfaces). When they are brought together
as a system, by activating certain (or all) of these abilities to interact, the properties of
the system emerge. The difference, when humans are brought together and interact
to form a society, is that the humans change as a result of the interaction. Not their
physical characteristics, but characteristics associated with behaviour, with how they
process information, as determined by the contents of the Memory. And so, while
we speak of an individual, and this individual's interaction with other individuals,
this individual exists only as part of a society. It is this ability of the individual to
change as a result of being part of a society that determines the dynamics of society,
and it makes the dynamics so much more complex than the dynamics of "classical"
engineering systems—systems with a fixed, predetermined behaviour. This same
complexity is now appearing in engineered systems with the ability to learn and to
act autonomously on the basis of the system's previous experience.

The model indicates that there are two types of incoming information. One is by
interaction with the environment in which we are embedded as members of society;
it is the information we receive by listening to the radio, watching TV, or reading the
paper or a book, conversing with other people, or just glancing at publicly displayed
advertising, an interaction with what we shall call the *circulating information*. It
is existing information presented to us, whereas the information labelled *discovery*
is obtained as a result of a targeted action by us, e.g. by doing an experiment or
by otherwise observing our environment, and results in an increase in the available
information.

It might be tempting to say that without discovery there would be no new information; the same information would just keep circulating between individuals. But that
is not true; it is a basic feature of a system that the system will display behaviours—
the so-called emergent behaviours—that are not displayed by any of the elements,

and in our case it means that combining information in the Processor can result in new information. In particular, even in the case of no incoming information, the Processor can work on the knowledge in the Memory; resolving contradictions and finding new correlations and relationships between items of information. We could think of it as "making sense" of the knowledge; a process that can lead to an abrupt change of the knowledge base and may also result in a need to provide the new-found information items as outputs to the rest of society. This process of reflection or analysis includes logical deduction, insight, and revelation; it is often identified as turning knowledge into understanding or into belief, but perhaps the best description of the function of this analytical or critical activity is maintaining a high degree of internal consistency in the Memory.

Modelling the individuals as identical information processors is suitable for many aspects of society and its processes, but in some cases we need to take account of the individuals as humans, and this immediately raises two issues. The first one is reproduction, which involves a particular interaction between male and female individuals based on biology (in addition to any exchange of information), and therefore does not seem to fit into our system description. However, we can sidestep this issue by replacing both reproduction and death by a renewal process within the collection of individuals making up society (or a group within society); it becomes part of what we might consider to be the *metabolism of society*. That introduces the second issue— the *lifecycle* of the individual. In our view of society as a system, the individuals are time-independent, and the evolution is an evolution of energy, from free to bound; the individuals are simply the carriers of the process. In the same vein, we can now consider the evolution of society to be an evolution of the knowledge and the inter-actions, with the physical beings constituting a background on which this evolution takes place. Under this perspective, the early education and integration of children into society becomes part of the metabolic process; the process of maintaining the living substance of society. It is somewhat similar to describing the behaviour of a person without considering the fact that, within that person, cells are dying and being created and integrated into the body at a great rate. The role of education in the dynamics of society will be an important subject further on.

Finally, we make a simplification in connection with the old controversy about nature versus nurture; that is, to what extent the behaviour and mental capabilities of a person are determined by inheritance. There are two aspects to this issue: One is that we all inherit traits characteristic of our species, such as processing of sensory inputs, memory management, reasoning, and speech. There is no doubt a certain variation about a mean value of these traits and capabilities, but whether they show any systematic variation (race, gender) or are completely random is irrelevant for us, as we operate only with averages. The second aspect concerns additional behavioural traits, such as a predisposition to physical aggression, or the interpretation of a dif-ference in looks or behaviour as a threat, or more ethical traits, such as truthfulness, altruism, and compassion. Are they inherited or are they a result of the environment in which a person grows up; i.e., of information received? If we, for a moment, use an analogy between the processor and a computer, then it is difficult to distinguish whether the variation in the output (behaviour) resulting from a certain input (situation) is due to a difference in the firmware (mental processing capability) or in the software (experience) the computer uses in order to produce the output. The simplification is to assume that the mental processing capability is the same for each person, and that a person's behaviour, including the extent to which it exercises the processing capability, is determined only by its experience, from which it follows that the content of the individual's memory is determined by the individual's interaction with society, and is something that has to be created anew in every individual, hence the crucial role of education. The interaction is with persons that represent the history of society over a preceding period equal to, in principle, the age of the oldest living person. This time, and the time it takes to create the initial memory content, are two irreducible time constants that enter into the dynamics of society.

3.3   The Interaction

The interaction between the individuals in a society was the main subject of *The Social Bond*, where it was treated in considerable detail and with an emphasis on binary interactions and the diffusion of information. Some of that material is repeated in this essay, but with a different focus—how it relates to the evolution of society. If we compare the interaction between individuals with the other component of the information processing system—the individuals, we find a very significant difference, and one that forces us to treat these two components differently. Within the time frame of interest to us—the last 10,000 years or so—the individual, as a means of processing information, has not changed significantly; only the information stored in and processed by the individual has changed. But in the case of the interaction, not only have the amount and content of the transmitted information changed, but the means of interconnection have changed at an increasing rate. For our purpose it is useful to divide this process of change into three phases, each with distinct characteristics. In the first phase, the transmission of information was essentially direct from person to person, mostly verbal, but also by hand signs and body language. The only storage of information was in the human memory, and the only spatial propagation of information was in the movement of humans.

The second phase is characterised by the encoding of information in various forms, such as signs, letters, drawings, paintings, and sculptures. This introduced a dependence on technology, in the form of tools for producing the encoding, and in the form of the media to contain and store the encoded message. With the advent of printing and photography the cost of production was reduced by orders of magnitude, but at an increasing dependence on technology. Throughout this phase it was the case that while the transmission was not restricted to being in real time and spatially within the reach of the human voice, the spatial transmission of information depended on the medium being physically transported, as by the postal service. The storage capacity was greatly expanded, as was the temporal aspect of the storage (we have clay tablets from more than 4000 years ago). This phase also introduced a novel feature: The possibility of anonymity and/or false attribution of authorship. Without face-to-face interaction, "the written word" could be given an authority that was unverifiable, ranging from divine to scientific, and "fake news" took a big step forward.

The third phase, which started with the invention of the telegraph, is the current phase of human interactions dominated by telecommunications and information technology. This technology includes landline telephone, radio, television, digital data transmission, such as the Internet, and mobile telephone, and in addition to only transmission of information, the technology provides the means for processing the information. Compared with the two preceding phases, the two important differences are that most of the communication is not a two-way interaction, but a one-way provision of content, often for the purposes of advertising or entertainment, and that the transmission does not involve the transport of a physical medium as the carrier of the information, which means a reduction in cost by orders of magnitude. The result is that information, in its various forms, has become a very significant aspect of modern life, and that what is shown as the flow of information in Fig. 3.1 is very much more complex than just a direct interchange between two individuals. As only an indication of the present role of IT-mediated information exchange in our society is the amount of time we spend on IT-based activities, including fixed and mobile voice and data, TV, and radio. There are various estimates, e.g., from eMarketer (www.emarketer.com/) and Hacker Noon (https://hackernoon.com), and while they are fairly poorly defined, it would appear that in the US people spend several hours per day watching TV and as much as 4 h per day on mobile devices, of which 1.5 h on social media and 2.5 h on apps. This is certainly considerably more than they spend on face-to-face interaction.

From what has been presented in this and the previous sections we see that, at any one point in time, the output of the information-processing, and thereby the actions that constitute the change in society, is determined by the content of the Memory. But the content of Memory is itself constantly undergoing change as a result of the information flow, which is in turn changing as a result of changes to the information technology. Thus, if we, in a very schematic manner, denote society by S, the memory by M, and the information technology by T, we have two interlinked processes: $dS/dt = f(M)$, and $dM/dt = g(T)$. The first one is the subject of the next chapter; the second one is treated in Chap. 5.

3.4 Identity and Belief

A great deal of the information in the individual's knowledge base, stored in the Memory component of the model of the individual illustrated in Fig. 3.1, is the basic information that allows us to function and survive in our environment; it is what children have to learn by experience, such as a flame being hot, water not providing a firm surface, and how to hold a pen. And there is information that we would classify as "facts", such as that the circumference of circle with radius r is $2\pi r$; these are information items that are simply used in carrying out mental tasks, but that require no re-evaluation. In what follows, we shall be particularly interested in a small subset, Θ, of the knowledge base contained in the individual's memory; the knowledge that determines the evolution of society by compelling us to take adaptive action, actions that are intended to result in a change to society. This subset contains statements about what a person believes in and is willing to sacrifice something for. It consists of information items of class 3 or, basically, information items that are subject to *persuasion*, and we shall call this subset the individual's *identity*. It is close to what is identified as *attitude* by Daniel O'Keefe in his book *Persuasion* (2002) and will be the meaning of 'identity' throughout this essay. The concept of the identity was elaborated in considerable detail in *The Social Bond*, so only a few central characteristics are repeated here.

The information items contained in the identity, which we shall call *identity items*, are basically statements (opinions, beliefs) with the following structure:

> I believe Sydney is the greatest city in the world because its building A is the tallest building in the world.

Here "Sydney is the greatest city in the world" is the *assertion* of the statement; "I believe" simply identifies the assertion as asserting a belief, not a fact, and as all identity elements express information beliefs, this part is unnecessary (or optional). The phrase "because its building A is the tallest building in the world" is the *argument in support of the assertion*. The assertion consists of two components: "Sydney" is the *subject* of the assertion, and "is the greatest city in the world" is the *predicate* of the assertion. The subject of the statement, in this case "Sydney", is one of the things we have called "the things that matter"; it is a thing about which I have beliefs or opinions, and about which I can make assertions. It is possible for an identity to contain assertions for which it has no arguments. The above structure of identity items can then be further detailed as follows:

> An assertion is composed of a subject, S, and a predicate, which we shall identify by the character σ, so that an assertion may be identified by S^σ. The set of all arguments in Θ associated with the assertion S^σ shall be denoted by $X(S^\sigma)$.

For a particular assertion, S^σ, there will be a large number of possible statements, each with a different argument. These arguments may not all be independent, in the sense that one argument may *entail* a number of other arguments. For example, the above statement entails all statements of the form "I believe Sydney is the greatest city in the world because its building A is taller than building B in Buenos Aires". Then, keeping in mind that the interaction between individuals is the basic process in our view of society as a system, we further narrow the definition of $X(S^\sigma)$ as containing all arguments in support of S^σ that would be recognised as such by the individuals involved in the interaction, but such that no argument entails any other of the arguments. The number of arguments in $X(S^\sigma)$ is the *strength* of the assertion, denoted by ν.

The definition of the identity as "the information items that matter" is a very high-level definition, but it will suffice for the purpose of this essay, and it avoids some controversial issues; in particular, the issue of "knowledge" versus "belief". If we use the word "belief" for an identity item, the assertion is what we believe, and the argument is why we believe it. The extent to which the argument is grounded in a scientific paradigm, in an ideology, or in religion, is not relevant to our use of the concept of an identity item, and the issue of "truth" does not arise.

Let w be the number of identity items in Θ. Some of these items are related, in that they have the same assertion; but differ in their arguments. Let $s(v)$ be the number of assertions appearing v times in Θ, with

$$\sum_{v=1}^{w} v \cdot s(v) = w, \tag{3.1}$$

then we can view $s(v)$ as a distribution of strength, characteristic of the identity. We shall make the simplifying assumption that the distribution can be approximated by a *uniform distribution*, defined by $s(v) = w/v$. The two extreme distributions are $s(w) = 1$, and $s(1) = w$. With this, it seems intuitive to define the *strength of an identity*, γ, by the expression

$$\gamma = \sum_{v=1, s \neq 0}^{w} \frac{v}{s(v)}; \tag{3.2}$$

with $1/w \leq \gamma \leq w$, and the relationship between the number of assertions in the identity, z, and the identity strength is then given by the expression

$$\gamma = \frac{w}{z^2}. \tag{3.3}$$

It may at first seem strange that the strength depends on the size of the identity, w, but if we consider the two extreme cases above, then, in the case of single assertion with w arguments we would expect the strength of the assertion to increase with the number of arguments supporting it, and in the case of all assertions having only one argument, the strength becomes increasingly 'diluted' with increasing value of w.

Thus, to an identity, Θ, with w identity items, there corresponds a *set* of assertions, ϑ, with z members, and in a society of n individuals we can form the union of these n sets,

$$Q = \bigcup_{i=1}^{n} \vartheta_i. \tag{3.4}$$

Each assertion in Q, q_i, $i = 1$ to $i_{\max} \leq n \cdot z$, is characterised by a parameter κ_i, its *commonality*, $1/n \leq \kappa \leq 1$, so that $n \cdot \kappa_i$ indicates the number of identities in which q_i appears. Associating commonality with assertions rather than with identity items is because what is of primary importance in a society is the commonality of what people believe; the extent to which they have the same reasons for believing it is of secondary importance.

Fig. 3.2 The distribution of the commonality, κ, of the assertions in the society. Approximating this distribution by a rectangle defines the value of the parameter α

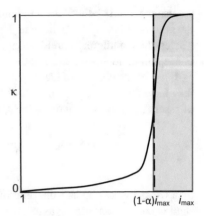

Now pick any one of the items in Q and determine its commonality, continue this process for all the i_{max} items, and then order them from left to right in order of increasing commonality; we obtain a curve like that shown in Fig. 3.2.

If we do the same for the items of an identity, we get a similar curve, except that the number of assertions along the x-axis is z instead of i_{max}. In particular, the number of assertions with κ equal to 1, or close to 1, will be the same in both cases. So, while α is a global measure of how closely the individuals are integrated into society, and a further expression of the degree of social integration, it also a measure of an individual, in that it defines a subset of Θ with $α \cdot z$ assertions. It is a characteristic of the identity, dividing the identity into two parts: the α-part contains the assertions common to all the individuals, what Durkheim calls the *conscience collective* [3, 38], and the $(1 − α)$-part contains the assertions particular to each individual (or to small subsets of individuals). We can therefore think of α as being one measure of *social cohesion*.

Implicit in our idealised, high-level model of society is the free flow of information between individuals; there are no other elements or structures within this uniform society that could influence the flow in any way. The flow is generated by the individual and, as far as the information items in class 3 are concerned, the flow is composed of the identities of every individual. The individual's beliefs only have a value in so far as they are made public; the importance of this public role of the individual was emphasized by Hanna Arendt in her discussion of *vita activa* as political action [4]. As a result, we can think of the class 3 elements in the flow as a *circulating collection of identity items*, with the total number of identity items in the circulating collection, W, given by $W = n \cdot w$, and within that collection a collection of Z assertions, with $Z = n \cdot z$. This means that the assertions in the α-part of Θ each appear n times in the circulating collection, and that α has the same meaning in the circulating collection as it does in Θ.

We shall call the α-part of Θ *society's belief system*. The significance of this characteristic in the context of our model will become clearer further on, and it will form the focus of our deliberations in Sects. 3.6 and 4.5, but it raises an issue about our use of averaging as a means of simplification. For any parameter measuring a characteristic of an individual or interaction the concept is straight-forward; we sum the values of the parameter for all individuals and divide by the number of individuals. An example is w, the size of Θ, which will vary greatly from individual to individual, but where the average still makes perfectly good sense (how useful or significant it is, is a different matter). But, as we realised above, the same is not true of the information being stored or processed by the individuals. When we speak of the individuals in our model as "average" individuals, it is as information-processing entities; ignoring the identity imparted to each individual by the content of Θ_i. A corollary to this is that the life cycle of a person becomes "invisible"; the "average" individual has properties that are averages over persons in all stages of their life cycles. So, for example, if no new information was generated within society, all the parameters of our model would settle into steady-state values, even though each person experiences great changes in its knowledge and information exchange with other persons throughout its life time.

Altogether, the concept of "the individual" as a representative member of society, which is central to our model of society, is only significant because the members, over which the average is formed, constitute a society; i.e., have a certain level of interaction. The result of this interaction is what is represented by the parameter α, but while α characterises both society and "the individual", the individual member of society has no knowledge of the value of α, nor of the value of κ for any of its identity elements. Hence, the decisions taken by an individual member cannot depend on α; what an individual member is aware of is the distribution of strength, $s(v)$, and the strength of the identity. This feature of our model is a result of neglecting the reinforcing effect of repetition as an explicit characteristic of interaction. The effect of repetition, well known in advertising, can be included in the persuasiveness of an interaction intended to effect a change in the identity, but the repeated input of an existing identity item is ignored, as will be shown presently.

The strength of the identity, γ, is a numerical parameter, and so we can form the average over all members of society and associate this value with the individual. We then have two parameters characterising the identity of the individual, Θ: α and γ, and it is reasonable to ask if there is a relationship between them. There does not appear to be a direct relationship, in the sense that the value of one determines the value of the other, which is not surprising, as α is a measure of the alignment *between individuals*; i.e., how similar their beliefs are, irrespective of what the beliefs are; whereas γ is a measure of how narrow, or focused, the belief *of the individual* is, irrespective of whether this focus is the same for different individuals. For example, consider two cases, both with $\alpha = 1/2$: In one case, $z = 2$, $v = w/2$, and $\gamma = w/4$; in the other case, $z = w$, $v = 1$, and $\gamma = 1/w$; two very different values of γ for the same value of α.

The identity is that part of the knowledge base the individual uses to evaluate information that relates to what the individual believes in and what determines any adaptive action, including the adaption of the identity itself, and the main evaluation criterion has always been survival. Society, with its organisation, institutions, and norms, is the result of this process of evaluation and adaption, and until recently the single most important input into the process was knowledge about the physical environment in which the society existed. Different environments resulted in different societies; an obvious example being the Aboriginal societies in Australia, where the absence of plants suitable for agriculture, and animals suitable for husbandry (at least after the megafauna became extinct), led to societies very much attached to the land as an entity they respected and with which they coexisted, generally without any desire to change it or any belief in their ability to do so (an exception to this is the eel farm discovered in Western Victoria). Following on from this example, we can identify the abundance of natural resources in the form of arable land and water, relative to the size of the population, as a main characteristic of the environment in which a society existed prior to the industrial revolution. And it is possible to see a correlation between this parameter and a parameter that characterises the role of the individual in the society, *freedom*, ranging from complete freedom, or individualism, with no obligation to society, to a highly structured society where each individual has a fixed position, purpose, and obligations. That is, essentially no personal freedom to move within, or to change, the structure. The greater the abundance of resources, the greater the individual freedom; there is enough to go around, and there is less of a requirement to be efficient. With more constrained circumstances, there is a greater need for cooperation and for overall efficiency, which favours a hierarchical organisation with more central control. A demonstration of this correlation that is relevant today arises from comparing the evolution of society in China with that in the West, the latter being in recent times epitomised by the US. Up until about 1500 AD, the main natural resource was land suitable for agriculture, in particular cropland, defined by the Food and Agriculture Organisation (FAO) as the sum of arable land and land under permanent crops, and while there is no data on cropland going back beyond 1700 AD, various estimates have been made of per capita cropland [5, 6]. From these estimates it appears that, while the value of per capita cropland has been steadily decreasing everywhere, the value for China was around half that for Europe. Food supply, both its production and distribution, was always a major issue in China, with famine an ever-present threat, so that a well-ordered, fairly rigid and hierarchical structure was accepted as the best. This desire for stability was the main reason behind the social philosophy set out by Kongzi (Confucius) around 500 BC, and it has been a characteristic of Chinese thought ever since. And it is probably fair to say that, up until 1800 AD, the Chinese political system was superior to any in the West, judged by the lifestyle of its citizens. The small farmers were not poor, and existed within a good social order with strong family values and equality between men and women, with the husband tending the fields and the wife weaving. And commerce was well ordered and not heavily taxed, if at all; with transport infrastructure in the form of roads and canals a government concern. Until at least 1800 AD, China was the most prosperous and productive nation in the world [7].

But it was also an inward-looking system that considered China as the civilised part of the world; the people outside were considered barbarians with whom closer contact, beyond commerce, was of no value and would at most lead to instability, and in the nineteenth century China's foreign trade represented only a tiny fraction of its huge internal commerce [7]. China had no desire to conquer or colonise any other part of the world, and even though Chinese technology was at least on a par with European technology in, say, 1450 AD, when ocean-going ships made exploration and acquisition of overseas land possible, the Chinese showed no interest in this, despite the expeditions of Zeng He.

Contrast this with the situation in Europe, which had been a cauldron of conquests and manoeuvrings between kingdoms, fiefdoms, and tribes at least since Roman times. The relative richness of the land provided scope for individual initiative, and so, when opportunity knocked, around 1450 D, in the form of vast new areas of the world opening up, the Europeans were ready to go out and conquer the lands, plunder their riches, and exterminate or enslave the local populations. It was the enormous wealth that flowed into Europe from this exploitation that financed the great surge in science and technology from the Renaissance onward and then, when the United States was born, fuelled the great industrial wealth created there and made the US into the dominant nation in the Western world. This abundance of resources to be had for the taking has led to a view of society as a competition with as few rules and restrictions as possible; a glorification of individual freedom as opposed to social solidarity, and an ideology that misunderstands, or overlooks, the actual nature of society and the role of the individual in it, and of how these are changing. As a result, there may be a small, but important difference between the identity of a Chinese and that of a Westerner, reflected in how the individual perceives what is the optimal balance between individual freedom and social solidarity, and it is a difference that may become increasingly important as more people want to compete for diminishing resources.

The idea that the circumstances that were present at the creation of a society are reflected in its further development was noted by de Tocqueville in his work on Democracy in America [8], where he compared it to the development of a person being determined by the circumstances of early childhood:

> ... we must watch the infant in its mother's arms; we must see the first images which the external world casts upon the dark mirror of his mind; the first occurrences which he witnesses; we must hear the first words which awaken the sleeping powers of thought, and stand by his earliest efforts, if we would understand the prejudices, the habits, and the passions which will rule his life. The entire man is, so to speak, to be seen in the cradle of the child. – The growth of nations presents something analogous to this: they all bear some marks of their origin; and the circumstances which accompanied their birth and contributed to their rise affect the whole term of their being. If we were able to go back to the elements of states, and to examine the oldest monuments of their history, I doubt not that we should discover the primal cause of the prejudices, the habits, the ruling passions, and, in short, of all that constitutes what is called the national character; we should then find the explanation of certain customs which now seem at variance with the prevailing manners; of such laws as conflict with established principles; and of such incoherent opinions as are here and there to be met with in society, like those fragments of broken chains which we sometimes see hanging from the vault of an edifice, and supporting nothing. This might explain the destinies of certain nations, which seem borne on by an unknown force to ends of which they themselves are ignorant." (Chap. 2, Part I)

He saw America as a prime example of this, as the creation of that nation took place within relatively recent times and is well documented, whereas most other nations emerged as amalgamations and divisions of numerous societies over a very long period of time and often poorly documented. China is an exception, in that despite great internal turmoil, the basic concept of China as a society and a nation has endured since its inception over 4000 years ago, providing some validity to the comparison of China and the US on the basis of their origins.

3.5 The Two Cycles

Central to the present development was the realisation that the activities that take place in society can be grouped into two cycles. One, the *economic cycle*, contains the greater part of the activities, they are the ones that sustain and grow the current "version" of society. The other, the *transformation cycle*, contains the activities that change one "version" into the next, and it is this process we understand by "the evolution of society". This grouping is reflected in dividing information into classes, as we did in Sect. 3.1, and with the individual's actions being determined by its identity. But this then immediately raises the question: what sort of change to society constitutes a transition into a new "version"? Or, in other words, how do we define a "version"?

Within our high-level model of society as an information-processing system, the answer is that a "version" of society is defined by its *belief system*, as we defined it in Sect. 3.4, and the activities of the transformation cycle are those that change the belief system. And so, within this model, the evolution of society is synonymous with the evolution of society's belief system, and it is the result of the activities of the transformation cycle.

In both cycles, the activities are of two types: the activities internal to the individual; i.e., the mental activities performed by the Processor in Fig. 3.1, and activities external to the individual; i.e., physical activities performed by the individual as interactions with its environment, and we have asserted, in Sect. 1.1, that the latter are completely determined by the former. The purpose of an external action, which reflects the preceding internal activity, can be to effect a physical change, to exchange information with other individuals, or a combination of both, but it may sometimes be difficult for an external observer to determine exactly what the purpose is. As a consequence, while in theory and for our model it is possible to distinguish between the activities of the two cycles, in practice this is not so, and it is one of the reasons why the division into two cycles is neither obvious nor generally accepted.

Another reason is that, for most people, the activities in the economic cycle dominate by orders of magnitude; they are the activities that make up our daily life and support our existence as physical beings. If we look at the measures used to compare societies, they are mainly such ones as current GDP, growth of GDP, education level, research expenditure, military power, etc.; all measures of activities within the economic cycle. If we consider how most of us spend our time, as we did briefly in Sect. 2.4, not much is spent on activities in the transformation cycle; at least, not consciously and productively. And if we look at the issues dominating the political agenda, particularly at election time, as was so starkly demonstrated in the most recent federal election in Australia (the nation with the greatest *per capita* wealth in the world), it is all about the economy; not a thought for what sort of a society we would want to be. Under the pressure of providing profitable investment opportunities for the ever-increasing capital, we are being turned into consuming robots.

This focus on the economic cycle was lamented some time ago by Marcuse [9]. Instead of two cycles he talked of two dimensions, and his second dimension is essentially the transformation cycle. For this process to function, there needs to be tension in the first place; it only becomes active when the individual perceives a conflict between the social environment in which it is embedded and its own identity. Acceptance of this conflict as a prerequisite to progress is central to Marcuse's critique of modern society. Referencing Hegel, he states that the power of the negative is the principle which governs the development of concepts, and contradiction becomes the distinguishing quality of reason [9, 171]. And concept is taken to designate the mental representation of something that is understood, comprehended, known as the result of a process of reflection [9, 105]. It is this dialectical mode of thought that Marcuse calls *negative thinking*. Negative thinking is the driver of Marcuse's second dimension of the development of society; the driver of the first dimension—*positive thinking*—is the acceptance of the perceived world as the basis for reason. The difference between the two modes of thinking is reflected in the difference between "is" and "ought" [9, 132].

Marcuse's concern is summarised in the subtitle to the Introduction to *One-Dimensional Man*—"The Paralysis of Criticism: Society without Opposition". And there should be criticism of a society which he describes as follows: "Its productivity is destructive of the free development of human needs and faculties, its peace maintained by the constant threat of war, its growth dependent on the repression of the real possibilities for pacifying the struggle for existence—individual, national, and international." His analysis of why, in the face of this situation, there is so little criticism, reflects his background as a philosopher sharing many roots with Heidegger, Adorno, and Horckheimer, but at a high level, and in a perhaps overly simplistic view, it is about information; how the control of information can determine our perception of reality and make the irrational seem perfectly rational. The current state of society is made to appear so obviously the logical choice that there is no need to look for any alternative; it's just the way it has to be.

The suppression of the transformation cycle, or the second dimension, is a result of a strong coupling between the two cycles, in that maintaining the status quo of the economic cycle and acceptance of it inequalities always required the suppression of the activities of the transformation cycle. The control of information is nothing new, and religion has always played a major role in effecting it; it was always a central aspect of the catholic church and was exercised with a degree of cruelty that made it a form of terror. The Renaissance was the beginning of a revolt against this control; a process that culminated in the Enlightenment. However, at this time the Industrial Revolution started to realise its own demand for control; a control required for the acceptance of the alienating and degrading conditions of the industrial production process. Today the control is required by the economic cycle in order to maintain a version of itself that has long passed its use-by date.

3.6   The Individual Process

Following on from the realisation that only a small part of the total information input to the individual is relevant to the evolution of society (which is our focus), we can expand the information processing shown in Fig. 3.1 by explicitly showing the processing of information of class 3; i.e. information relevant to the identity.

This is shown in Fig. 3.3, where we identify two processes.

Process A is the main process, which evaluates the inputs being received from other individuals and from the individual's own discovery. The former include not only interactions in real time, such as face-to-face, or via some electronic medium, but also reading (inputs from the author), viewing movies, TV, etc., where inputs are provided by script writers, directors, advertising designers, and the like. The latter may be from direct visual observation, but increasingly via instruments of varying complexity, and which have allowed us to realise that what can be directly observed via our senses is only a very limited view of Nature. In general, the evaluation has one of four outcomes: One is the discarding of any information item that is already in memory. Two is the updating of the knowledge base by incorporating some of the received items and discarding some existing ones; three results in taking action in the form of producing output items that either support or oppose received items, and the fourth is the identification of information items that conflict with any in Θ; in this case Process A simply transfers them to Θ for resolution by Process B. By "conflict

Fig. 3.3 The two parts of the processing function, A: evaluating the inputs, B: the individual's background processing (contemplation) of the contents of the knowledge base, and the two parts of the Memory: the identity, Θ, and the rest (general memory)

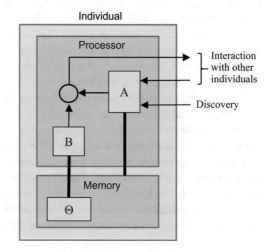

with" we shall understand not only direct conflict, as in negation, but 'question', 'suggest modification', or 'extend'.

Process B is the individual's reflection on the contents of Θ, the "making sense" of the knowledge; a process that can lead to an abrupt change of the identity and may also result in a need to provide the new-found information items as outputs to the rest of society. This process includes logical deduction, insight, and revelation; it is often identified as turning knowledge into understanding or into belief, but perhaps the best description of the function of Process B is maintaining a high degree of internal consistency in Θ, as was noted earlier.

If we consider Process A to be an evaluation process, then the result of the evaluation of inputs in class 1 is that no action of any form is required. Evaluation of inputs in class 2 may result in a judgement in the form of a probability distribution, but this uncertainty is a reflection on the input, not on the evaluation criteria. In both cases, no doubt is raised about the existing evaluation criteria, whereas evaluation of inputs in class 3 does raise such doubts. Process B is also an evaluation process, but of the evaluation criteria themselves, in light of the doubts raised by Process A about inputs in class 3. From the definition of activity cycles in the previous chapter, we see that Process A is primarily involved in the economic cycle, whereas Process B is concerned with the transformation cycle; with the critical thinking Marcuse called negative thinking.

If we accept that the inherent processing capability of the individual, measured in information items per unit time, has remained unchanged over the last 10,000 years, the change in the value of the actual processing capacity is determined mainly by the change in the fraction of time available for this activity (i.e., waking hours), but compared to the increase in the flow of information and the growth of knowledge, this increase is irrelevant, and we shall take this capacity to be a constant, and label it as u_0. The issue is then how this processing capacity is allocated to the two processes, A and B, and within Process A, to the three classes of the input. Addressing the latter first, this is where the concept of *attention*, introduced in Sect. 3.1, comes in. If I sit on my balcony, relaxing and looking out over the Pacific, or gazing out of the bus window, or scanning a newspaper, or being entertained by a TV show, or surfing the Internet, there is a vast amount of information impinging on my sensory system, but it is immediately identified as requiring no further action; it is all class 1 information. The processing capacity used for this activity depends of the richness of the environment in which I find myself: how fast it is changing, how detailed it is, whether it is just visual or involving other senses, and so on; two extremes might be gazing out over the ocean and standing in Times Square. And even if I close my eyes, Process A continues to process old information from memory. But if some item of information catches my attention, by being unexpected and identified as requiring further processing, the processing of the class 1 information is greatly reduced. The extent to which it is reduced depends on the person's ability to concentrate, and in most cases it continues to operate as a background process that we identify as distraction.

Now, it appears that the power consumed by the brain is about 12 W, and that this power is relatively constant and independent of any particular brain activity, see e.g. [10]. We shall therefore assume that the energy required to process an informa-tion item is proportional to the time required, and interpret the components of the processing capacity as the fraction of the total time available for mental processing. This total is, essentially, the waking hours, but with a reduction for some activities that occupy us completely and prevent us from carrying out any effective mental processing, such as certain types of manual labour, so that, for the time being, we might surmise that this total is in the range 12–15 h. In our model of the individual we account for this by saying that the individual spends a component u_1 of its pro-cessing capacity on class 1 information, whether exclusively or as background, and a component u_2 on processing class 2 information. A further component, u_3, is spent by Process B on processing those information items that Process A has identified as being in class 3. In any case, for our individual, the sum $u_0 = u_1 + u_2 + u_3$ is a fixed quantity, and the dimension of u_i is hours per day.

The question is then: 'what determines the values of these components?', and we approach an answer to this question by making a few observations. Firstly, we recognise that processing class 1 information requires significantly less energy than processing class 2 information, so that there is an inherent tendency to classify input as being in class 1, and we classify information elements as being in class 2, i.e., as being worthy of our attention, only if they promise to bring us some benefit. Constructing the information so that it is more likely to be perceived as class 2 information rather than class 1 is, of course, the purpose of marketing or, more generally, of persuasion. But there is the more basic aspect of this, in the form of the information processing associated with work and the benefit of receiving payment, and while the fraction of time spent at work is diminishing, the increase in the ratio of intellectual to manual work is putting upward pressure on u_2. Secondly, the increasing complexity of society and of life in this society is resulting in a need to process a growing amount of class 2 information (as in bureaucracy and "red tape"). Thirdly, the information arriving at Process A from the environment (i.e., through discovery) is inherently in class 2.

The picture we have of the individual's information processing so far is then as follows: In what we might consider the default mode, Process A first processes the input to separate out the class 1 information from the class 2 information, taking no further action with the class 1 information, except perhaps store some of it as fleeting impressions. With the class 2 information, which has captured the individ-ual's attention, Process A switches to a focused and detailed mode, resulting in new information being produced and stored and/or transmitted. During this processing, some of the information is identified as being class 3, and stored in the section of the memory we have labelled Θ. Process B now recognises that there are inconsistencies in Θ, and uses a fraction, u_3, of the processing capacity to reduce or eliminate them. The change in the level of unresolved inconsistencies thus depends on the difference between the flow of class 3 information and the rate of resolution by Process B, and in the next section we shall suggest that this level of unresolved conflicts, to be denoted by β, can be identified with Fletcher's behavioural stress.

A two-process model of the cognitive function has been around since the work of Amos Tversky and Daniel Kahneman, as reviewed in Kahneman's Nobel Prize lecture [11]. It is suggestive to identify what they call System 2, Reasoning, with

Process B, and their System 1, Intuition and Perception, with Process A (refer to Fig. 3.1 on p. 451). (Other authors, many of which are referenced in [11], have called System 1 and System 2 slow and fast thinking, respectively.) However, this identification is only partially correct, as System 2 incorporates also the processing of the class 3 inputs to Process A, and, as is discussed in the next section, it misses an essential feature of Process B—the creation of new information.

To investigate the dynamics of the information processing we have described, we need to introduce the rate at which information arrives at an individual—the *information flow*. In both Figs. 3.1 and 3.3 we identified two inputs: One was from other individuals in the form of *interactions*, and we shall label this flow of information by μ_a, in units of identity items per unit time. The other, in the form of *discovery*, was from the individual's own exploration of the environment, and we shall label this flow of information by μ_b. Only a part of the information received by an individual is received through direct face-to-face interaction with another individual; most of the information is received through various information channels, as already noted. Supporting these channels is a significant part of society's infrastructure, based on information technology. Applications of this technology are involved in processing, distributing, and, above all in the present context, storage of information. As a result, there is a vast pool of information, which we identify as the *collective memory*. This information was, of course, all produced by individuals in the course of time, and while it is subject to a certain amount of attrition (items being lost), the increase in this store of human knowledge is rapidly accelerating. It is true that the overwhelming majority of this increase is related to science and technology, but there is also an increase in what we called class 3 information, much of it extending the applicability of previous understanding to the complexity of the current society.

Let us then define the portion of μ_a which catches the attention of the individual as $x \cdot \mu_a$. Of this information, only a small part, say y, will relate to the information items in Θ; i.e., be of class 3. We can then introduce the parameter μ_1 to represent the flow of information relating to Θ, i.e. $\mu_1 = y \cdot x \cdot \mu_a$; this is the flow which is evaluated by Process B, and when we now try to understand what determines this flow, and how this influences the identity, we need to keep the following points in mind:

i.   The individual has no knowledge of society's belief system, and will evaluate the inputs and act accordingly based solely on its own identity. However, there may be a connection between the two, as in the case where an individual says "I don't really believe in this, but I will do it anyway, because I believe others will approve"; that belief is part of the identity.

ii. The information items arriving through the interaction are a random selection of the items in the circulating information, and the items in the flow μ_1 are a random selection of the identity items in the circulating collection of identity items, W.

iii. We use the rectangular approximation, as illustrated in Fig. 3.2, to define the composition of the circulating collection of identity items.

iv. In the interaction with the circulating collection of identity items, the size of the
     identity, w, remains constant. We can observe, in ourselves and in others, that
     we can only maintain a limited number of things that matter to us, that we are
     passionate about, that we are willing to sacrifice something for. If we accept new
     items, develop new passions, they tend to displace some existing ones. There
     is certainly a significant variation between people in the number of causes they
     care about, but we shall assume that, from adolescence onward, which is the part
     of the lifetime in which the individual can influence the evolution of society,
     the size of the individual's identity is fixed and equal to w. That is, new identity
     items *replace* existing ones.

Note that the size of the identity is measured in terms of identity items rather
than in terms of assertions. The reason for this is that a strong belief tends to replace
weaker ones; in the extreme case of a belief with strength w there is no room for any
other belief.

The assertion associated with an item arriving as part of μ_1 is either identical to
one already in Θ, and the probability of this is α, or it is different to any assertion in
Θ, and the probability of this is $(1 - \alpha)$. In the first case, the item is either identical
to one in Θ (i.e., has the same argument), in which case it is simply ignored, or it
provides a new argument for the assertion, which may or may not be accepted. If it is
accepted, the value of ν for that assertion increases by 1 and that of another assertion
decreases by 1. However, at the current, high-level development of the model this
process, which results in a change of the strength of the identity, is ignored, and the
value of ν is the same for all assertions (the uniform distribution discussed earlier).
In other words, the current model is concerned with changing the belief system in
the sense of replacing some beliefs by new ones, not with strengthening an existing
belief. The latter can be viewed as a separate process; essentially the realisation of
the first commandment.

In the second case, the assertion is potentially in conflict with an existing assertion
in the $(1 - \alpha)$-part of Θ (by definition, there can be no conflict with the assertions
in the α-part), and needs to be processed by Process B, so that there is a flow of
potentially conflicting items, $(1 - \alpha) \cdot \mu_1$. We shall now introduce the concept of the
stress arising from the level of unresolved conflicts and denote it by β, measured as
a fraction of w, and then express the change in this level as follows:

$$w \frac{d\beta}{dt} = \mu_1 \cdot (1 - \alpha) - \mu_2 \cdot \beta; \tag{3.5}$$

where $\mu_2 \cdot \beta$ is the rate at which Process B resolves conflicts; this defines μ_2. That
this rate increases with increasing value of β is reasonable; as the stress increases, we
put more effort into lowering it. But setting the rate proportional to β is somewhat
arbitrary and done just for simplicity, and it ignores any saturation of the capacity of
Process B.

As far as the identity, Θ, is concerned, the resolution of a conflict has, basically,
one of two outcomes: Either the conflicting item of information is accepted, which
deletes one item from the $(1 - \alpha)$-part of Θ and adds one to the α-part, *assigning the*

same value of ν to it as that of the deleted item (extending the definition of ν to be either the size of the set X or a value transferred in a replacement), or the conflicting item is rejected. Which of these two outcomes eventuates depends on two things: the *persuasiveness*, p of the conflicting item, and the strength of the assertion in the item to be replaced, ν. The probability of replacement becomes lower the greater the strength, and in the absence of any other information, we may model this by the function $1/\nu$. The persuasiveness, which is a characteristic of the argument, shall be the probability of acceptance when $\nu = 1$, so that the probability of replacement is p/ν. The resulting change to α can be expressed as a function of p and ν:

$$\Delta\alpha = \frac{p}{\nu \cdot z}. \tag{3.6}$$

Furthermore, as previously mentioned, each individual generates new information, both as a result of discovery and as a result of reflection or analysis. The former was shown as the discovery path in Fig. 3.1 and labelled by μ_b above; the latter, which may be considered an 'internal' flow, shall be labelled by μ_c. This combined flow of new information is important in determining the longer-term evolution of society, in particular if α approaches 1, but it is always true that both μ_b and μ_c are much smaller than μ_a; our regular information processing is concerned overwhelmingly with issues we are familiar with, rather than on new discoveries. Of course, only a small amount, μ_0, of the information entering through this channel is of class 3 and results in a change in Θ. If the item to be replaced is in the $(1 - \alpha)$-part, there is no change to α, but if it is in the α-part, it is deleted and the replacement added to the $(1 - \alpha)$-part. (Note that we are assuming that there is no conflict involved in this replacement process.) Hence, the rate at which α is decreased by this "rejuvenation" process is $\mu_0 \cdot \alpha$, and we obtain

$$z\frac{d\alpha}{dt} = \mu_2 \cdot \frac{p}{\nu} \cdot \beta - \mu_0 \cdot \alpha. \tag{3.7}$$

The two external and two internal sources of information, and the three classes of information, can be summarised in Table 3.1.

The two equations, Eqs. 3.5 and 3.7, form a set of coupled linear first degree differential equations for the two functions $\alpha(t)$ and $\beta(t)$. Solutions to these equations

Table 3.1 Categorisation of the information flows by source and class

Source		Class		
		1	2	3
Interaction	μ_a	$(1 - x)\,\mu_a$	$x\,\mu_a$	$xy\,\mu_a \equiv \mu_1$
Discovery	μ_b		$(\mu_b + \mu_c)(1 - \mu_0)$	μ_0
Analysis	μ_c			
Resolution	μ_2			μ_2

are well known, but without knowledge of the values of the coefficients it is not possible to determine a particular solution. But we can solve the two equations for the steady state, resulting in the following expressions:

$$\alpha = \frac{\pi}{\pi + \omega};$$ (3.8)

and

$$\beta = \frac{\omega}{\pi + \omega} \frac{\mu_1}{\mu_2};$$ (3.9)

where $\pi = p/\nu$ and $\omega = \mu_0/\mu_1$. Equation 3.8 shows that if $p = 0$, then $\alpha = 0$, which means that the society has disintegrated, and it is "each man for himself". That is, a certain amount of persuasiveness, or, what is more appropriate in terms of π, a certain amount of acceptance is required in order to establish the relationship, or bond, between individuals that forms them into a society. But we also see that it is the quantity ω that prevents the alignment from becoming absolute ($\alpha = 1$); it is what Big Brother tries to suppress—a measure of *individuality*. The function $\alpha(\pi; \omega)$ is displayed in Fig. 3.4.

To recognise the most important feature of the function $\alpha(\pi; \omega)$, as displayed in Fig. 3.4, we have to realise that, in a modern society, $\omega \ll 1$; the individual (i.e. the average person) receives the overwhelming majority of information items in Θ from society. And for a normal, or healthy, society, we would expect to find the value of α to be somewhere around 0.5. In this case, the value of α is very sensitive to the value of π, which obviously has a significant implication for the stability of society. We shall come back to this after we have considered what characteristics of society determine the value of π.

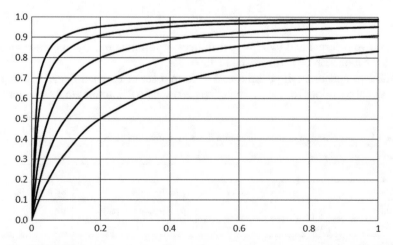

Fig. 3.4 The function $\alpha(\pi; \omega)$ for five values of the quantity ω, from top to bottom 0.01, 0.02, 0.05, 0.1, and 0.2

The stress level, β, is the proportion of items in Θ that are being challenged by new information originating from other individuals. But as only items in the $(1 - \alpha)$-part of Θ can be challenged (as there is full agreement on the items in the α-part), we have $\beta \leq (1 - \alpha)$, with the equality meaning that the identity of the individual is completely challenged. A condition which, if persisting, would lead to something like insanity, and that we might possibly, in the next section, equate to Fletcher's I-limit. As a consequence, we have the condition

$$\frac{\mu_1}{\mu_2} \leq 1; \tag{3.10}$$

which basically says that the flow of information that challenges the identity of an individual must be less than $(1 - \alpha)$ times the capability of the individual to resolve such challenges. (This should be a reminder of Ashby's Law of Requisite Variety [12].)

Equation 3.9 shows that the level of conflict within Θ, which we have equated to stress for the individual, depends on the various flows of information. But can these flows take on any value? The total time available to the individual for processing these flows is u_0, and to investigate how this time is distributed on the three components of information flow we identified earlier, we can describe their time consumption by the following equations:

$$u_1 = a_1 \cdot \mu_a; \tag{3.11}$$

$$u_2 = a_2 \cdot (x \cdot \mu_a + \mu_b + \mu_c) \leq u_0 - u_1; \text{ and} \tag{3.12}$$

$$u_3 = a_3 \cdot \mu_2 \cdot \beta = a_3 \mu_1 \frac{\omega}{\pi + \omega}; \tag{3.13}$$

where a_i is the time spent on processing one item of information in component i. Equation 3.11 represents a simplification, in that we are taking this component to be a constant background process, depending only on the amount of circulating information, and not on its composition, as already indicated.

The inequality in Eq. 3.12 implies the order in which the three components can draw on the total processing time: first u_1, then u_2, and then u_3, if there is any time left. It also shows that there is an upper limit on x:

$$x \leq \frac{u_0 - a_1 \mu_a - a_2(\mu_b + \mu_c)}{a_2 \mu_a}. \tag{3.14}$$

This upper limit turns the individual's attention into the limiting factor in the consumption of information; in the words of Matthew Crawford, "Attention is a resource—a person has only so much of it" [13]. And Eq. 3.14 displays the expected dependence on $(\mu_b + \mu_c)$: the more the individual is immersed in its own projects (e.g. study and research), the lower the available amount of attention that can be allocated to the circulating information.

What we have developed so far is a model of the state of society as an information-processing entity, characterised by the alignment, α, and the stress level, β. They are determined by the information flows and a particular, and highly simplified, view of the individual who processes these flows, which introduced a large number of parameters about which we have little or no quantitative data. But if we recall one purpose of this development, as stated at the beginning of the Introduction, we may not need this quantitative information in order to determine if there is a stability problem or not, and we shall return to this issue in Chap. 8. What is easily discernible from our model is the important role of technology: Increasing the input through interaction by applications of communications technology, increasing the proportion and persuasiveness of class 2 content through data processing, and increasing the input through discovery using improved technology (instrumentation, access to data). In short, our view of society as an information processing entity is dominated by the flow of information—its quantity and its composition, and this flow is determined by the applications of technology involved, as we explore in Chap. 5.

Before leaving this section, we need to realise that the development of the process taking place in the individual's identity as a result of the interaction with the circulating information contains an apparent contradiction. On the one hand, we use the rectangular approximation, which means that the commonality, κ, is either $1/w$ or 1, and the identity is divided sharply into two parts by the parameter α. On the other hand, when an individual accepts an assertion from the circulating flow, the commonality increases from $1/w$ to $2/w$. The resolution of this apparent contradiction is that the individual is the representative of all the members of society, so that what happens to the individual happens to all members, and the value of κ does go from $1/w$ to 1. However, what we have suppressed is the fact that this does not take place instantaneously; the process of how κ of an information item goes from $1/w$ to 1 is a process characteristic of the society, not of the individual, and will be taken up in Sect. 4.5.

3.7 Stress and the Development of Settlements

If we look back in time, the only practical means of transmitting information was by word of mouth, and the determining factors were population density and mobility. Here we can draw on the seminal, and for us highly relevant, work of Roland Fletcher in studying numerous settlements, both ancient and more recent [14]. In this work, Fletcher describes how the growth trajectories of settlements depended on the relationships between residential density, settlement size, and what he calls "the material assemblage" available to the occupants (which I identify as the extent of technology and its applications), and this growth was constrained by two limits: the upper limit on a tolerable residential density, determined by the stress experienced by an increasing level of interaction between the individuals within a settlement

(I-limit), and the limit on the settlement's areal extension, set by the distance over which the prevailing communication system could operate adequately (C-limit). The interaction consist of all the sensory inputs experienced by a person due to the other people in the settlement, including through such channels as sound, body language (i.e. through vision), odours, and physical means (e.g. crowding, bumping into), as defined in (ibid. 70). Communication is the exchange of messages between people, both face-to-face and mediated by various applications of technology (message tokens, clay tablets, etc.), and where the face-to-face exchange could be arranged through movement of the persons involved, so that means of person transport (roads, bridges, horses, carriages, boats, and more recently mechanised transport) became part of the communications technology. It is interesting to note that although goods transport uses much the same infrastructure as people transport, it is not included as a factor in determining the C-limit (ibid. 82).

From Fletcher's analysis of the great amount of data collected by himself and other authors, two significant realisations emerge. The first is that, for a settlement to function as a cohesive entity, there needs to be a certain level of information exchange across the settlement, and therefore, for a given communications technology (in the above sense) there is a definite limit to the areal extension of a settlement. The minimum size at which the benefits of a sedentary versus a nomadic lifestyle could be demonstrated seems to have been on the order of 10^4 m^2. These agricultural settlements grew until they reached a limit of 10^6 m^2, with communications provided mainly by walking. This limit was only exceeded once the technology for building compact urban settlements, in the form of retaining walls, drainage, roads and bridges, was developed and applied, together with various forms of writing, and such settlements developed rapidly in size up to a limit of 10^8 m^2. This limit was maintained, with a few exceptions, until thermal power started its spectacular development at the beginning of the Industrial Revolution in the mid-eighteenth century; first the steam engine and then the internal combustion engine, supporting a revolution in personal transport and in printing. This was accompanied by such technology developments as electricity and telecommunications, and industrial cities grew in size to the next limit, 10^{10} m^2, which brings us to the present.

The second realisation is that there is an upper limit on the level of population density, and that this level has remained essentially constant throughout the evolution of settlements. Equating this limit to a limit on interaction (behavioural) stress, it says that despite all the changes to the environment in which interpersonal interaction has taken place (the built environment), the effect of this interaction (i.e., the stress and its limit) has remained unchanged. That is, this limit is a fundamental characteristic of the human psyche, of the human as a social being.

Fletcher describes the development of settlements in terms of three variables:

p population
a area of the settlement
d density of population within the settlement

If we, at least to start with, take the density to be uniform within the settlement, we have the relation $a = p/d$, as shown in Fig. 3.5, and the growth of a settlement is

Fig. 3.5 Lines of constant area in the p-d plane, with $a_1 < a_2 < a_3$, and a possible trajectory of the development of a settlement

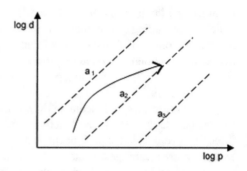

represented by a trajectory in the *p-d* plane. The stress, *s*, experienced by a person due to the presence of another person, must decrease as a function of the distance, *r*, between the two persons, and let us assume a simple, exponential decrease,

$$s = s_0 e^{-\frac{r}{\sigma}}. \tag{3.15}$$

We have chosen *r* as the symbol for the distance because we now consider the situation where the person experiencing the stress is in the middle of a settlement within a circular area of radius r_0. In this case, the total stress experienced by this person, S, in units of s_0, is given by

$$S = 2px^{-2}\left(1 - e^{-x}(x + 1)\right); \tag{3.16}$$

where $x = \left(\frac{p}{\pi \sigma^2 d}\right)^{\frac{1}{2}}$. The value of σ is determined by requiring that S is the same for the two endpoints of the I-limit in Fig. 4.5 of [14]:

$$p = 10; \quad d = 0.0642 \text{ persons/m}^2;$$
$$p = 10^8; \quad d = 0.0294 \text{ persons/m}^2;$$

from which it follows that $\sigma = 4.55$ and the critical value of S (the I-limit) is $S^* = 3.83$. The value of S/S^* as a function of p and d is shown in Fig. 3.6.

This result shows what we would expect—behavioural stress is a local phenomenon; once the population reaches about 30 people, further growth at constant density does not increase the stress. However, with growth at constant density, the area of the settlement increases proportionally, and the communication between members becomes increasingly difficult and inefficient. But it is this communication that turns the members of the settlement into a society; it is what underpins the fabric of society, and so this decrease in communications efficiency represents an increasing stress of a different type—a stress on the fabric of society. The two types of stress are not independent, or orthogonal, as the frustrations of a poor and overcrowded public transport system demonstrates. The behavioural stress experienced by a person depends on the homogeneity of the person's environment; people

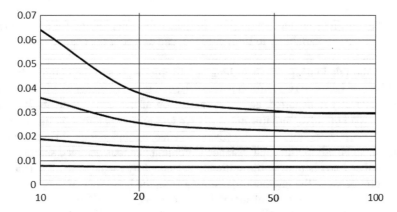

Fig. 3.6 Lines of constant behavioural stress experienced by persons in a settlement, as a function of population, p (horizontal axis), and density, d (vertical axis, in persons/m^2), as a fraction of the critical stress value, S^*—the I-limit in [14]—with, from upper to lower curve, $S/S^* = 1.00, 0.75, 0.50$, and 0.25

who are visibly different (hair style, clothing, skin colour) or who behave differently (language, gestures, body language) increase the stress. This effect is reduced if the settlement is segmented into racial and/or socio-economic areas, but this also reduces the communication across the settlement, so that there is a certain conflict between reducing the behavioural stress and increasing the social cohesion. A similar effect may take place once the density level is reached. The settlement needs to restructure itself and expand into a larger area, thereby lowering the density, before starting on another period of increase, resulting in a saw-tooth history of population density. With it, there is likely to be a certain structuring into local communities (boroughs, arrondissements, Bezirke), so that our assumption of a uniform density is only an approximation, and the sense of belonging to a local community reduces the behavioural stress, while probably also reducing the cohesion within the settlement as a whole.

Nevertheless, it is both significant and useful to distinguish the two types of stress, and we shall relate them to our high-level view, which is concerned with society as a whole. However, before doing that, there are a couple of issues we have to consider. The majority of the settlements on which Fletcher's analysis is based existed prior to the Industrial Revolution, which gave these settlements two features. One, the developments in communications technology that permitted a transition across a C-limit were preceded by developments in the activities that sustained the settlements, first agriculture, and then, forms of extraction and manufacturing technology, in particular tools and building materials (bricks, concrete, metals, chemicals). The technology was developed during the expansion of the settlement and, when this expansion came up against the C-limit, this technology was available to be applied to the improvement in communications required to expand the settlement size across the C-limit. This, as well as the interdependence of the interaction and the communication, was discussed in (ibid. Chap. 6). This does not detract from the importance of communications technology in providing the cohesion required for a settlement, but adds a certain variability to timing, speed, and extent of the changes in different settlements.

Two, the relationship between the area of settlements and the communications facilities was between two entities parametrised in terms of distance. Messages had to be physically transported, and so an extension of the area automatically meant a further load on the communications system and increased stress on the ability to maintain social cohesion. With the advent of telecommunications, this relationship changed, and the importance of distance in communications declined sharply.

A further issue is that, in the settlements studied by Fletcher, there was a clear distinction between the interaction, resulting from the behaviour of people, and the communication, as a means of maintaining coherence of activities across the settlement. The latter was largely an administrative function, and while the population would be very aware of its effects, as e.g. in taxation, and of the stress resulting if it was inadequate, as in increased lawlessness, it was something that the great majority of the population would simply accept as given. In a modern society, that is quite different, with an increasing proportion of the stress experienced by individuals arising from concerns about the state of the society and from responding to the demands of government at all levels, and an increasing proportion of the interpersonal, or behavioural, stress occurring via the communications infrastructure. The distinction between the two types of stress—on individuals and on society—is becoming less clear.

A final issue is that of structure; both of the settlement and of the society. As already mentioned, as the complexity of the society increased, there could be a segmentation of the physical space into racial and/or socio-economic groupings, and also in that people with the same interests and beliefs congregate within subareas of the settlement. A further source of spatial segmentation is the structuring associated with the division of labour, both hierarchically (levels of administration) and by trade (e.g. "tinsmith alley"). As a settlement grew in size and complexity, this spatial segmentation reached a point where the society could be considered as consisting of several settlements. This issue of spatial structure is not one that is relevant to our high-level model, but it is one that will play a significant role in the future development of settlements, as outlined in the final chapter of [14]. However, in the present age—the information age—there is a different type of structuring that is important, as discussed below.

To see the significance of Fletcher's work in the context of our model, we need to relate his concepts of "interaction" and "communication" to the features of the model; i.e., Processes A and B, and the knowledge base Θ, as shown in Fig. 3.2. The "interaction" does not feature explicitly in the model, but we now identify the stress resulting from the interaction as resulting from an unacceptable degree of inconsistency within the knowledge base, Θ, and it is also experienced as a need for certainty or for closure. This is a well-known and -documented feature of the human psyche; Googling "need for certainty" gives more than a million results. The effect of inconsistency among the elements of Θ is the subject of a special field within cognitive science called cognitive dissonance theory, based on the idea that persons seek to minimise the internal psychological inconsistency of their cognitions (which equate to our identity items—the elements of Θ). Cognitive inconsistency is taken to be an uncomfortable state, and hence people are seen as striving to avoid or minimise it. A good description of cognitive dissonance theory is given in Chap. 4 of Daniel O'Keefe's book *Persuasion* [15].

In our model the stress arises when the time available for executing Process B is inadequate in relation to the flow of new information being placed in Θ by Process A. This implies that there is an upper limit on the information processing capacity available to, and shared by, the two evaluation processes, and that, if the flow of information is great enough, Process A will tend to use up the available time, crowding out Process B.

Fletcher's "communications" consists of information that is in classes 2 and 3. In class 2 we find the information of an administrative nature, associated with the various processes that keep a modern society functioning, such as restrictions (laws) of various sorts, taxation, licensing, and reporting ("red tape"). In class 3 we find the information that maintains a certain minimum value of the alignment α throughout the society—the characteristics of the society, its culture. Therefore, if the communications infrastructure is stretched beyond its capacity, there are two separate consequences. One, the services that provide the benefits that constitute the rational for a society in the first place start to deteriorate. Two, the common belief system that supports cooperation disintegrates. Both of these can be viewed as exerting a stress on the fabric of society that either limits further expansion or, possibly, continues expansion as two or more societies.

References

1. Aslaksen EW (2018) The social bond: how the interaction between individuals drives the evolution of society. Springer, Berlin
2. Axelrod R (1997) The dissemination of culture: a model with local convergence and global polarization. J Confl Resolut 41(2):203–226
3. Durkheim E (1984) The division of labor in society (trans: Halls WD). The Free Press, New York
4. Arendt H (1958) The human condition. Chicago University Press, Chicago
5. Pongratz J, Rick C, Raddatz T, Claussen M (2008) A construction of global agricultural areas and land cover for the last millennium. Glob Biochem Cycles 22(3). Published by the American Geophysical Union
6. Goldewijk KK, Dekker SC, van Zanden JL (2017) Per-capita estimations of long-term historical land use and the consequences for global change research. J Land Use Sci 12(5):313–337
7. Bankoff G, Boomgard P (eds) (2007) A history of natural resources in Asia: the wealth of nature. Palgrave Macmillan, London
8. de Tocquevile A (1835) De la démocratie an Amérique. English edition: de Tocquevile A (1838) Democracy in America (trans: H. Reeve). Allard and Saunders, New York
9. Marcuse H (1964) One-dimensional man. Routledge & Kegan Paul Ltd., Abingdon
10. Jabr F (2012) Does thinking really hard burn more calories? Scientific American, 18 July. Available at https://www.scientificamerican.com/article/thinking-hard-calories/
11. Kahneman D (2002) Maps of bounded rationality. Nobel Prize Lecture, available at http://nobelprize.org/nobel_prizes/economics/laureates/2002/kahneman-lecture.html

12. Ashby WR (1956) An introduction to cybernetics. Chapman & Hall Ltd., London
13. Crawford MB (2015) The world beyond your head: on becoming an individual in an age of distraction. Farrar, Straus and Giroux, New York
14. Fletcher R (1995) The limits of settlement growth. Cambridge University Press, Cambridge
15. O'Keefe DJ (2002) Persuasion: theory and research, 2nd edn. Sage Publications, Inc., Thousand Oaks

Chapter 4
The Collective Intelligence

Abstract The evolution of society is presented as an ongoing process, without any other purpose that its own survival. This survival is threatened by fluctuations, analogous to temperature fluctuations in a physical system, but managed by the operation of the collective intelligence—the integrated operation of the individual processors, with the common part of their identities forming society's belief system.

4.1 The Meaning of Survival

When we speak of "survival", it means on the one hand that something persists in time; on the other hand that it does so in spite of something that is detrimental to its persistence. As individuals we speak of surviving an ordeal or a challenge, but when we speak of the survival of the human race, the species *homo sapiens*, or human *society*, it must relate to something other than the survival of individuals, which is severely limited. On the level of the individual in any species, the ability to assess if a change in the environment is good (beneficial to survival) or bad (a threat to survival) is a defining characteristic of what we call *life*, and is readily observed in the various life forms around us, as well as in ourselves. And it is evident that as the complexity of the environment increases, the complexity of the criteria for assessment and associated actions increases accordingly.

If we think of interacting humans as forming a new life form, we can see the emergence and disappearance of various societies over the last 10,000 years as the evolution of this life form with the societies as the species, as was already suggested in the Introduction. And, in analogy with the evolution of life from a single cell to *homo sapiens*, there has been an evolution of society from small groups of hunter-gatherers to today's world society. In both cases the process has been one of trial-and-error and selection based on the fitness to succeed in a competitive environment. The difference lies in the trial process—the process of generating new species. In the case of life forms, this appears to have been by means of largely random mutations and sexual reproduction; in the case of society the process has clearly been driven by us.

The understanding of what constitutes survival, as the criterion against which the assessment is made, evolves as society evolves. This is very much in the same spirit

E. W. Aslaksen, *The Stability of Society*, Lecture Notes in Networks and Systems 113, https://doi.org/10.1007/978-3-030-40226-6_4

as what Philip Kitcher argues in his book, *The Ethics Project* [1]. He demonstrates how our concepts of ethics and of what is ethical must, if they are to be relevant and effective, relate to our society, the environment in which our actions take place, and as this is a continually developing environment, the same must be the case for ethics. The issue is not primarily about ethical truth, but about ethical progress. Ethical truth is not preordained and something towards which we are striving: instead, we should be striving for each progressive change to be truthful. A similar view was expressed by Joseph F. Fletcher some time ago in his book *Situation Ethics*, where he stated "the morality of an act is a function of the state of the system at the time it is performed" [2]. In our model, the evolution of society is not towards any fixed vision of society; what is important is the *continuity and stability* of the evolution. What survives is the evolution of society.

4.2   The Importance of Fluctuations

So, how did we, despite many wrong turns, guide society from the cave to where it is today? How did we assess if a change to society was going to improve or reduce its chances of survival? The ability to do this is a hallmark of our species—the core of what we call *intelligence*—and it has two distinguishing features. One is the above understanding of what constitutes survival. The other is that it depends on the individual's knowledge; i.e., on what we have identified as its identity, Θ, and as a result, the outcome of the assessment varies from individual to individual, and from time to time. For this assessment to be effective on a society scale, there needs to be a process of forming a consensus, or an average, in the sense of a *collective intelligence*, and the enabler of this process is the free exchange of information between individuals.

To develop our understanding of collective intelligence, it is useful to start with a simple picture or model, as shown in Fig. 4.1. On the right is a box labelled "Recurrent activities"; they comprise the majority of the activities making up Life and of the economic cycle. They are the combined actions of the individuals making up the society, but as these individuals display more or less different behaviours, the collective behaviour is fluctuating. The box on the left, labelled "Collective intelligence", observes these activities and, in particular, assesses the fluctuations and takes adaptive actions to either promote them as changes to society or suppress them as undesirable; these actions constitute the transformation cycle.

Fig. 4.1 Life, as a combination of recurrent activities and actions driven by the collective intelligence

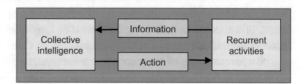

The concept of fluctuations in the state of society (however we measure it) is central to this understanding of the dynamics of society. Fluctuations are well known to us from daily life, ranging from fluctuations in daily temperature to those of the share market, but a particularly well studied case is that of fluctuations of the energy in a container filled with a gas. A molecule in the gas has a kinetic energy u, given by its mass, m, and speed, v, as $u = mv^2/2$, and the energy of the whole volume of gas, containing n molecules, is the sum of the individual energies,

$$U = \frac{1}{2}m \sum_{i=1}^{n} v_1^2 \tag{4.1}$$

The molecules collide both with the walls of the container and with each other, but between collisions the speed of a molecule is constant and given by a distribution. That is, there is a probability of the speed of a molecule being in the range between v and $v + dv$ that is given by f(v)dv, with f(v) being a well-known distribution (the Maxwell distribution). The average energy is $\bar{u}$, and the distribution is illustrated in Fig. 4.2 as the curve for $n = 1$.

Now, let us do the following thought-experiment: Consider a volume within the container that is so small that at any time there is just one molecule passing through it. Then, clearly, the energy in this volume will vary between zero and a high value, with a probability distribution as shown in Fig. 4.2 for $n = 1$. Increase the size of the volume so that there is, at any point in time, exactly two molecules passing through it. The energy of the gas in this volume is now the sum of the energies of the two

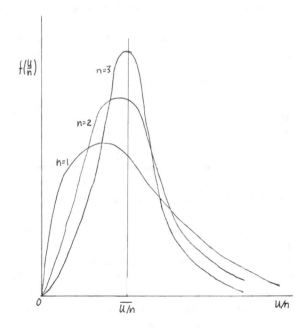

Fig. 4.2 The distribution of energy, U, within a volume of gas containing 1, 2, or 3 molecules (illustration only; more a work of art than scientific truth)

molecules, and the distribution is as shown in Fig. 4.2 for $n = 2$, as it is unlikely that both molecules will have a very low energy or a very high energy. Increasing the volume further, so that $n = 3$, the distribution becomes even more peaked around the average value, U, and as n increases, the distribution becomes narrower and narrower, with a half-width, σ, which we might think of as the average fluctuation, equal to $U/\sqrt{n}$. That is, the energy of the average fluctuation relative to the total energy equals $1/\sqrt{n}$.

To see how this result can be reflected in the evolution of society, we recall the view of evolution as a conversion of free energy into bound energy, with the ratio of free energy to total energy a possible measure of this evolution, or perhaps a component of a composite measure. Then the relative size of the fluctuations in the evolution of society is clearly a function of the number of members of the society. At the lower end of the progression shown in Table 1.2—the family—fluctuations are large, often resulting in the disruption of the society (e.g. divorce, family feuds, etc.), whereas at the upper end—the nation—fluctuations are less severe, and even revolutions (e.g. French, Russian, Iranian) seldom lead to the destruction of the society (but more on this later).

The validity of the analogy between this feature of a gas and of society rests on the issue of the effectiveness of the interaction in establishing an *equilibrium* state. In our view of society as an information-processing system, this means an unimpeded flow of information throughout the society. It is this flow, coupled with the individual processing, that results in the averaging, or consensus-building, that is one of the main features of the collective intelligence. Any interference with the free flow of information results in a sectioning of society into subsets of its members that, due to their smaller membership, are more likely to display large fluctuations from the equilibrium state.

The collective intelligence is the process that guides the evolution of society, and looking back at the history of this evolution, the fluctuations can be considered to be deviations from an ideal path, resulting from interference with the process, as illustrated in Fig. 4.3. Our understanding of the nature of this ideal path and of the fluctuations will be developed as we progress through this essay, and the first step is a highly idealised picture of the collective intelligence. Pick any individual, identified by a_i, with $1 \leq i \leq n$, and arrange the other individuals in the society on a plane grid, as shown in Fig. 4.4a, such that the intensity of their communications with a_i decreases with the distance from a_i, as shown in Fig. 4.4b. For simplicity, we approximate the intensity function as a rectangle, as indicated also in Fig. 4.4b; within the corresponding rectangular area in Fig. 4.4a there are m individuals, in addition to individual a_i.

Given this flow of information input, and a level of processing determined by the situation in which individual a finds itself (leisure time, health, education, etc.), individual a_i will produce an amount of new information and will also modify its own identity at a certain rate, as detailed in Sect. 3.6. That is, society will change at a certain rate and in a certain direction due to the information processing by individual a_i, and if this picture is the same for every individual in the society, which implies that each individual will appear within the rectangle in n/m pictures, the evolution

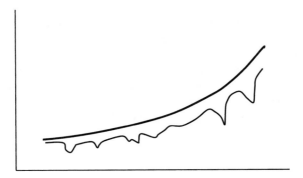

Fig. 4.3 The evolution of society (on a yet-to-be-defined measure), with the heavy, smooth curve representing the ideal evolution resulting from an undisturbed functioning of the collective intelligence, and the irregular curve representing the actual development, with the difference resulting from interference with the operation of the collective intelligence

Fig. 4.4 a Having picked one individual, a_i, shown in red, the other individuals in the society are arranged on a grid pattern, with their intensity of interaction with a decreasing with increasing distance. **b** The interaction intensity function, and its approximation by a rectangular function

(a)

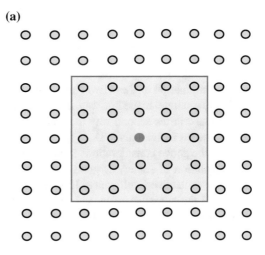

(b)

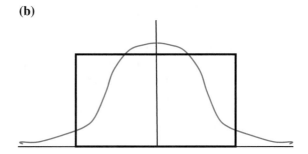

of society will be the ideal path in Fig. 4.3. This ideal evolution can be interfered with in two different ways: One, the situation in which individual a_i finds itself may reduce its processing capability, thereby reducing its influence on the evolution of society. Two, an individual appears in more than n/m pictures; in the extreme case of total thought control, the same group of m individuals appears in every picture. That is, if we think of the operation of the collective intelligence as an averaging process, determining the evolution of society by means of an averaging process of the form

$$A = \sum_{i=1}^{n} a_i; \qquad (4.2)$$

this process is changed in two principally different ways: In the first way, the contributions being summed are not equal, so $A \neq n \cdot a$. In the second way, the individual contributions are given a weight,

$$A = \frac{m}{n} \sum_{i=1}^{n} x_i a_i; \qquad (4.3)$$

where x_i is the number of pictures in which a_i appears. In the real world it will always be a combination of these two ways, but we shall find it useful to treat them as conceptually separate. In the first way, the change is due to a characteristic intrinsic to the individuals; in the second way the change is due to an interference with the interaction between individuals.

Implicit in the above considerations it what is the central premise of the present work: The evolution of society is good to the extent that it is determined by the exercise of the collective intelligence. It is, essentially, an affirmation of our trust in ourselves, in the human species, and with this, The Good Society becomes the society that provides the conditions for the exercise of the collective intelligence.

4.3   Society's Belief System

Consider what you are doing right now, reading this text. The Processor in Fig. 3.1 compares the information input with relevant material in Memory, discards some as already known, not interesting, or "clearly wrong", updates Memory with some, and takes an action, such as writing a note, in response to some. Then you might let your mind drift for a while, recalling some stored information or just listening to background noises, and finally you respond to an internal signal by going into the kitchen and finishing that piece of cake. None of this creates a conflict with your beliefs, and does not require a change to the identity, Θ. But a small part of the information we receive conflicts with the information stored in our identity and, as discussed in Sect. 3.6, the resolution, through Process B, of such conflicts may cause a change to the content of the identity. As a result, not only may we feel compelled to

Fig. 4.5 Distribution of commonality, κ, of society's assertions (or identity items, in this case)

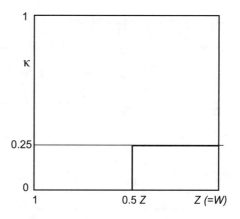

communicate this new insight to others, but our future behaviour will be changed, if only slightly. This individual process of resolving conflicts coupled with the interaction between individuals forms and maintains a degree of consensus we introduced in Sect. 3.4 as *society's belief system*; represented by the set of common identity elements—the α-part of Θ. But this concept entails more than what is expressed by the parameter α, and to get a feel for this, let us look at a couple of examples.

The first example is the following (extreme) society: Its members, n in total, are split into two groups, each with $n/2$ members. All the members have the same size identity, w, the assertions are all different, so that $z = w$ and $\gamma = 1/w$, $\alpha = 0.5$ in both groups, but the two groups have no identity element in common. For this society, $W = n \cdot w$, the strength, $\gamma = 1/w$, and the commonality distribution, as defined it Sect. 3.4, for the society, is as shown in Fig. 4.5.

There are no assertions that are common to all the individuals, there is no social belief system, as we have defined it, and α has no meaning for this society. What has happened here is the separation of the society into two *segments*, where a segment is a group of members that have an identifiable difference to the rest of society's members. In the case shown in Fig. 4.5 the segmentation is extreme, or complete, and we would consider the society to have disintegrated into two societies. But it does not have to be that extreme; the segmentation can be partial, leaving some identity elements common to all the members (i.e., with $\kappa = 1$). Our high-level model of society assumes that there is *no* segmentation; the individuals have no characteristics that would allow them to be distinguished; and α is the common parameter characterising both the identity and the social belief system. This is what is expressed by approximating the commonality distributions by rectangular distributions, as we did in Fig. 3.2.

The second example is a society without segmentation, as per our model, with the size of the identity, w, equal to 100 (in order to be more specific), and again with $\alpha = 0.5$. This society can have a range of strengths, depending on how the strengths of the assertions are distributed over the two parts of the identity—the α-part and the $(1 - \alpha)$-part, but in each part, the assertions all have the same strength. Some of the possible combinations are shown in Table 4.1.

Table 4.1 Some possible configurations of a society with $w = 100$ and $\alpha = 0.5$

z	$v_a + v_b$	v_a	v_b	γ
100	2	1	1	0.01
50	4	3	1	0.14
		2	2	0.04
20	10	9	1	1.00
		5	5	0.25
10	20	19	1	4.00
		10	10	1.00
2	100	99	1	100
		50	50	25

What Table 4.1 illustrates is, firstly, that for a given size of the identity, w, our definition of strength is a measure of the extent to which arguments are associated with a small number of assertions. Thus, a very high value of strength could also be characterised as fanatical or single-minded, and could be the result of indoctrination or brain-washing. Whereas a very low value of strength could be characterised as indecisive, shallow, or uncommitted. As with α, an ideal value is probably somewhere in the middle, say, $\gamma = 1$. Secondly, because γ is symmetric with regard to v_a and v_b, the strength does not discriminate between strong commitment to social values and solidarity, or to individualism. In using the model to investigate the dynamics of society, in Sect. 4.5, we shall ignore any details of the distribution of v, and assume that all the assertions in the identity have the same strength; i.e., a uniform distribution, as we already encountered in Sect. 3.4.

In our model of society, the collective intelligence is the information-processing activity resulting from the interaction of the processing ability of the individual members of society, and as such its nature is that of a consensus-building process. The actions resulting from the operation of the collective intelligence depend on society's belief system; those identity items form the basis on which Process A evaluates the information input. If we recall the definition of an identity item, in Sect. 3.4, it consists of an assertion and an argument in support of that assertion, and while the assertion expresses a belief, the argument relies to a significant extent on information that has more the character of data or understanding than belief. For example, if I believe that the operation of the collective intelligence is the most important function in society, I still need to understand what factors influence that operation and in what manner they do so, in order to evaluate the input information and decide on the action that will be most beneficial in supporting or improving the collective intelligence. This is largely a matter of what we might call *factual knowledge*, it is information that is *presented* as being factual, although the *truth* of this information is a different matter.

The factual information was, at some stage, received and then updated at some rate through the same input channel as the information to be evaluated (which we characterised by the parameter μ), and we can see that this offers a feed-back loop, or a process for positive reinforcement. By modifying factual data the evaluation criteria are modified, making the acceptance of arguments supporting a particular assertion more likely, which further modifies the identity, and so on. The effects of this feed-back feature of the collective intelligence, both good and bad, are well known (e.g. as in the phrase "sowing the seeds of doubt"), here we address the issue of which characteristics of society influence the operation of the collective intelligence. Such a set of characteristics will, of course, evolve as society evolves; the relevance of individual characteristics is only relative to a particular society within a particular period in time, as we noted in Sect. 4.1. In *The Social Bond* the issue of what characteristics of society influence the operation of the collective intelligence was approached by asking "what restrains its operation", and the following five restraints were identified (for more details, see Sect. 6.1 of *The Social Bond*):

a. *Financial restraint.* Let a person's yearly income be q, the subsistence level at the location where the person lives be s, and the average income per person be q_a, then the financial restraint, r_1, is given by

$$r_1 = e^{-\frac{q-s}{q_a}} ; \tag{4.4}$$

where $q \geq s$, and $r_1(q < s) = 1$. The interpretation of this restraint is that at (or below) the subsistence level the individual has essentially no time or energy to exercise intelligence; all effort is absorbed by mere existence. As the income rises above the subsistence level, the restraint is reduced, but the reduction depends not directly on the value of this increase above the subsistence level, but on the relationship of this value to the value of the average income in the society.

b. *Educational restraint.* For evaluating changes to society, the lower levels of education are considered more important, literacy is very much more important than having a Ph.D., and the *attained educational level* is weighted as 0.6 for primary education, 0.9 for secondary education, an 1.0 for tertiary education. The educational restraint, r_2, is simply

$$r_2 = 1 - \text{attained educational level.} \tag{4.5}$$

c. *Information restraint.* The ability to obtain information consists of two separate factors: *Access* to the media carrying information, and the *quality* of the media. If both of these are measured on a scale of 0–1, the information restraint, r_3, is defined by

$$r_3 = 1 - \text{access} \cdot \text{quality} \tag{4.6}$$

d. *Process restraint.* Being able to formulate an adaptive action is one thing; being able to action it is a different matter, and so lack of access to participation in a democratic and political decision process becomes a further restraint. Let u be the level of political rights, v the level of civil liberties, and w the level of democracy, all measured on a scale of 0–1, the process restraint, r_4, is defined as follows:

$$r_4 = 1 - (u + v + w)/3. \tag{4.7}$$

e. *Corruption restraint*. Corruption, in its various forms and at various levels within society, acts as a restraint on the operation of the collective intelligence in that it undermines the social attitude and effectively introduces a segmentation on the personal level. The corresponding restraint, r_5, is again measured on a scale of 0–1, with 0 being very clean and 1 being very corrupt:

$$r_5 = \text{level of corruption}. \tag{4.8}$$

Information about the values of these restraints is more or less readily available (see *The Social Bond* for details). How they are weighted and combined to form a single decision criterion will vary from person to person, but the average, as a characteristic of the individual and of society, becomes both an expression of the state of society and a determinant of the evolution of society's belief system.

4.4   The Basic Belief

A society will only survive if the understanding of what it takes to survive and its realisation in the structure of society (laws, institutions) develop at least as fast as the changes resulting from our development and application of technology. Or, in terms of our model, if the transformation cycle leads the economic cycle (i.e., is proactive), rather than lags behind it (i.e., is retroactive). In the past, it has been a case of responding retroactively, as was discussed in *The Social Bond*. There are two related processes: changes to society from applications of technology and the associated issues arising from these applications, and the process of developing restrictions to resolve these issues, and in a qualitative manner, we can illustrate their relationship as shown in Fig. 4.6.

The vertical distance between the curves, ε, shows the issues outstanding and needing to be resolved, and we see that despite the time required to resolve issues, Δ, is decreasing, it is not decreasing fast enough to prevent the number of outstanding issues, ε, to increase. In short, society's processes for responding to change are inadequate, and we seem to be facing a potential run-away situation. This problem was identified quite some time ago by William F. Ogburn in his book *Social Change with respect to culture and original nature* [3]. He considered the adjustment that was required between different parts of culture (or society, in our current terminology); in particular; the adjustment of the non-material culture to changes in the material culture, where the latter included both the natural environment and industry and its products, and showed that there was generally a delay between the change and the required adjustment, which he called *the hypothesis of cultural lag*. He discussed the

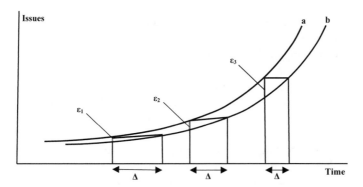

Fig. 4.6 The curve labelled **a** shows the issues that have arisen from the introduction of technology-based applications prior to any point in time, and the curve labelled **b** shows the issues that have been resolved. From *The Social Bond*

reasons for this lag, and in the Summary of Part IV, he considered an outcome which we shall also take up in Chap. 8:

> It is thinkable that the piling up of these cultural lags may reach such a point that they may be changed in a somewhat wholesale fashion. In such a case, the word revolution probably describes what happens. There may be other limiting factors to such a course of development, and our analysis is not sufficiently comprehensive and accurate to make definitive prediction. But certain trends at the present time seem unmistakable.

We note here the recurrent idea of two aspects to human existence—the physical and the mental—that appear in various guises in discussions of society. We have encountered it in Ogburn, in Marcuse, in the concept of the two cycles; it is reflected in C. P. Snow's *The Two Cultures* [4], and we shall encounter it again in the next chapter, in Durkheim's view of education.

There is a fairly general recognition of this problem and of the increasing urgency of finding and implementing solutions; there is much less agreement about how to go about it. And to a large extent this lack of agreement is a reflection of the issue we have encountered in several places in this essay so far: The individual's understanding of the nature of society and of the balance between the personal and the social. The lag, Δ, between the two curves shown in Fig. 4.6, is the time it takes for society's belief system—the α-part of the individual's identity—to adjust to the changes in society brought about by new implementations of technology, which in our model is represented by new knowledge. This increase in the knowledge needs to be processed by the collective intelligence and, in our view of evolution, result in a change in our understanding of the importance of the individual relative to that of society and in the associated process of determining the direction of the evolution of society. The main process for transmitting information is education, the topic of Chap. 6, and it is essential that this process is able to include new experience, insights, and understanding as they occur. Due to the stress involved in changes to an individual's identity, as discussed earlier, there is an inherent tendency to suppress information that is evidence for needed change, and one means of doing this is to ensure that it is not transmitted to the next generation—partially replacing education with indoctrination.

The understanding of the nature of society and the process driving its evolution put forward in this essay is based on the belief that the best responses to the changes arising from the economic cycle are provided by the operation of the collective intelligence; it is the ultimate realisation of democracy. With the realisation that all changes taking place within society are the results of thought-processes—evaluation of information, it seems reasonable to expect that, in the long run, the most stable evolution, the one most likely to avoid large, or even catastrophic, fluctuations, will result from having as many individuals as possible contributing. Conversely, the threats to the stability of the evolution of society are any actions, institutions, or structures that restrict this collective intelligence.

4.5 System Dynamics

The previous sections explored various aspects of the collective intelligence and of society's belief system, but how does the collective intelligence actually change the belief system? That is, how does the commonality, κ, of an identity item go from $1/n$ to n? It is clear that our model, as it has been presented so far, needs to be extended in order to be able to account for the dynamics of the belief system, as it does not allow for any other values of κ than $1/n$ and 1. In effect, our model presents a static picture of society, and an approximate one at that, as we have employed the rectangular approximation in order to give a simple meaning to the parameter α. This picture is useful for describing the state of a society, or for comparing societies, at any particular point in time; for a description of dynamics, the picture we need to have is something like what is shown in Fig. 4.7.

The flows of identity items indicated in Fig. 4.7 are as follows:

a New items created by individuals as a result of the operation of Process B, as discussed in Sect. 3.6.
b Items discarded by individuals as a result of the operation of Process B.
c The movement of items towards the boundary of groups 1 and 2, resulting from the interchange of information between individuals.
d Items that stop progressing in the process of increasing their commonality (i.e. ultimately unsuccessful persuasion), and end up either with $\kappa \ll 1$ or being discarded altogether (b above).
e Items, having reached $\kappa = 1$, becoming members of society's belief system.
f Items discarded or replaced from society's belief system.

In Fig. 4.7, the different identity items that exist in the society are arranged along the horizontal axis, and the vertical axis shows how many there are of each type. In group 1, the items are unique, $\kappa = 1/n$; in group 2 there are $n \cdot \kappa$ items of each type, and in group 3 there are n items of each type, so that the total number of items circulating in the society is the sum of all the elements in these three groups, $n \cdot w$.

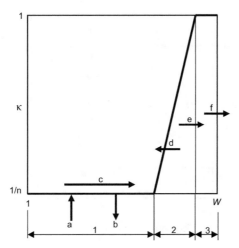

Fig. 4.7 The distribution of society's identity items according to their commonality, κ. The new group of items, 2, contains those items with a commonality less than 1, but that are, on the average, moving towards the right, i.e. increasing their commonality. Group 3 contains, as before, the α · w items of society's belief system, and the items in group 1 all have κ = 1/n. The arrows labelled by lower-case letters are flows of identity items

In order to develop an understanding of the process illustrated by Fig. 4.7, we start out, as usual, with a high-level picture of the process that changes the identity by accepting assertions from the stream of assertions circulating in society, for the case that all assertions have $v = 1$. The probability that a particular item of the set $n \cdot w \cdot (1 - \alpha)$ in the circulating stream will be accepted by a particular individual within one unit of time is given by

$$P = \frac{\mu_1 p}{w};$$

$$(4.9)$$

where p is the *persuasiveness*, introduced in Sect. 3.6. The probability that it will be accepted into any one of the $n - 1$ identities is $(n - 1) \cdot P$; and the length of time until it is accepted, t_1, is equal to $1/(n - 1) \cdot P$, and for this item, κ is now equal to $2/n$. The probability that one of these two identical items will be accepted by one of the $(n - 2)$ identities that have not yet accepted this item is $2P \cdot (n - 2)$, and we see that $t_i = 1/(i \cdot P \cdot (n - i)$. Then, after $n - 1$ time intervals, the assertion will have achieved κ = 1, and the total time, $T(n)$, is given by

$$T(n) = \frac{w}{\mu_1 p} \sum_{i=1}^{n-1} \frac{n}{i(n-i)} \equiv \frac{w}{\mu_1 p} F(n)$$

$$(4.10)$$

where $F(n)$ is approximated to a high degree of accuracy by the function $1.039 + 4.633 \cdot \log(n)$, as shown in Fig. 4.8.

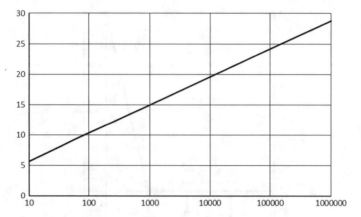

Fig. 4.8 The function F(n) (and its approximation; the two are not distinguishable)

Now, we do not have any data on the values of w and μ_1. A crude guess (and that is all it is) is that the number of statements of belief stored by an individual at any one time, i.e. the number of items, w, in the identity, is around 100, and that the number of information items relating to the identity arriving as input per day (this fixes the unit of time) is about 1. We are likely to reject most of these inputs, so p is small number, say, 0.01, so that the time for an assertion to rise from $\kappa = 1/n$ to $\kappa = 1$, i.e. to be included in society's belief system is on the order of $30 \cdot$ F(n) years. Leaving the numerical value aside for the moment, what might at first be surprising is the modest increase in F(n) with increasing n. The reason for this is our assumption that each individual that accepts the information item s into its identity also makes this fact public; this is the presence of the i in the denominator of the sum in Eq. (4.10). This assumption is simply a restatement of the basic assumption of our model of the individual and its role in society, as it was stated in Sect. 3.4 and developed in Sect. 3.6: The identity of the individual is reflected in society's belief system. We shall get back to this in Chap. 8; here we just note that if this public role is diminished, the value of F(n) increases; it takes longer to effect a change to society's belief system.

The model also shows the expected change in the rate of growth of κ, as illustrated in Fig. 4.9. Getting the first converts is always hard work; after that the growth is rapid, with an increasing number of believers in the public domain, but converting those last few unbelievers is not easy. (A similar behaviour is displayed by the well-known Bass Diffusion Model [5].)

This picture of the progression of the commonality of an identity item is highly simplified; in effect, it is a picture of the items in the flow labelled e. in Fig. 4.7, neglecting the process that reduces the value of the commonality on its way upward (stream d in Fig. 4.7) and greatly increases the time to reach $\kappa = 1$. The reality is that in a society of identical individuals, and in which the level of persuasion is the same for all assertions, it is highly unlikely that society's belief system will ever change, or that, starting with $\alpha = 0$, that a belief system would ever develop. That

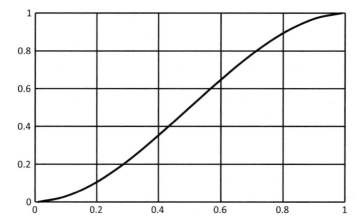

Fig. 4.9 The growth over time of the commonality, κ, for a single identity element. The time scale is in fractions of T(n) in Eq. (4.10)

a belief system does, in fact, develop in every society is due to the early years of the individual's life cycle, in which the exposure is not to a circulating information of uniform persuasion, but to a highly selected set of information with a level of persuasion close to 1, in the form of upbringing and education. This emphasises the importance of education, which is the subject of the next chapter.

The only significance of the process shown in Figs. 4.8 and 4.9 is as an indication of the growth of an assertion with a much higher level of persuasion than all others in the circulating information. A further complication of the process will result from increasing the value of ν above 1—increasing the strength of the belief system, this will reduce the rate of change, but the details of this effect are complex and well beyond the level of detail of this model. Replacing an existing identity item will be less likely; in Sect. 3.6 we had suggested a 1/ν dependence. Then, in order to replace one assertion, or belief, with another, there has to be ν individual replacements of identity items, resulting in one new identity item which, although it has only a single argument, has an assigned value of ν equal to that of the replaced assertion, as described in Sect. 3.6. We might expect that each successive replacement becomes easier for two reasons: One, the belief to be replaced is successively weakened as its arguments are being deleted, and, two, the new belief is becoming increasingly acceptable as arguments are being added. However, the latter effect depends on the rate at which the arguments are being presented; if the operation of Process B in accepting one argument for the new belief is still fresh in the memory, the acceptance of a second argument for the same belief is more likely—a fact well known in effective persuasion techniques. This process will presumably take more time the greater the value of ν, and there is likely an increase in the stress associated with it, as there is a prolonged period during which both conflicting beliefs are present.

At this point, the development of the view and model has been progressed about as far as intended for this essay, and before going on to consider some of the processes that constitute the environment in which the model operates, and then finally its relevance to the stability of society, it might be useful to briefly recapitulate the main features of the model. The model represents society as an *information-processing system*, consisting of *individuals* as the distributed processors and the *interaction* between them in the form of information exchange. The information is subdivided into three *classes*, of which class 1 is essentially a background for the other two, with class 2 the information of everyday life, and class 3 our beliefs. The information processing takes place in a physical environment consisting of *belongings* reflecting the state of technology, and the actions resulting from the processing mirror the two classes of information, with the actions resulting from class 2 information constituting the *economic cycle*, and those resulting from class 3 information constituting the *transformation cycle*. The central assertion of the model is that the processing of the class 3 information—the *collective intelligence*—and the transformation cycle determine the evolution of society, whereas the class 2 information and the economic cycle determine the environment in which the evolution takes place. That is, the collective intelligence has the dual role of maintaining society's equilibrium (or controlling fluctuations) and of guiding its evolution. This subdivision of information and associated action is useful, but its limitations are obvious.

References

1. Kitcher P (2011) The ethical project. Harvard University Press, Cambridge
2. Fletcher JF (1966) Situation ethics. Westminster John Knox Press, Louisville
3. Ogburn WF (1922) Social change with respect to culture and original nature. Huebsch Inc., New York
4. Snow CP (1959) The two cultures and the scientific revolution. Cambridge University Press, London
5. Bass F (1969) A new product growth for model consumer durables. Manage Sci 15(5):215–227

Chapter 5
Technology

Abstract This chapter considers one of the main components of the environment in which the information-processing takes place—technology and its application—and information technology (IT) is obviously of central importance. The main effects of IT are identified, as are its extent and economic significance in today's society.

5.1 Technology—And Its Applications

The word "technology" is used very frequently in our society; it is implicit in "modern", but its meaning is highly context dependent, and often quite broad and not well defined. This situation has been discussed in previous publications, and, as already noted in Sect. 2.2, we shall adopt the following definition:

> Technology is the resource base developed and applied by engineers in order to meet needs expressed by groups or all of society, and consists of a construction base (construction elements, tools, as well as the facilities within industry for fabricating and constructing plant) and a knowledge base (textbooks, publications, standards, data bases, heuristics, etc.). Students study technology in order to become engineers.

There is also frequently confusion regarding the difference between science and engineering, resulting in such expressions as "techno-scientist" and "technological sciences", obscuring the fact that the two have quite different purposes. Science is about truth; engineering is about being useful. However, science provides the foundation for much of the knowledge base of engineering and so, from the point of view of education, engineers need to have considerable knowledge of the results of science, and the development of technology is, to a great extent, driven by advances in science (although there is also much new technology arising from engineering itself). Conversely, applications of technology provide much of the infrastructure supporting science.

The origin of this confusion may be traced to the period following the First World War; a period of great advance in both science and engineering, and of a growing interest in social and philosophical questions relating to both. However, while the activity related to science was international in character, the activity related to engineering was concentrated in Germany and published in German (an important

E. W. Aslaksen, *The Stability of Society*, Lecture Notes in Networks and Systems 113, https://doi.org/10.1007/978-3-030-40226-6_5

example is the book *Streit um die Technik*, by Dessauer [1], first published under the title *Philosophie der Technik* in 1926), and as a result, when the interest in the philosophy of science and of engineering started to pick up under US influence following the Second World War, the philosophy of engineering, such as it was (and has largely been), adopted much from the philosophy of science, while disguising this by talking about "technology". A number of papers discussing and demonstrating this situation can be found in the publications resulting from the Forum on Philosophy, Engineering, and Technology, fPET 2012 [2], fPET 2014 [3], and fPET 2016 [4].

In the context of this essay, perhaps the most important issue with regard to understanding the role of engineering is the difference in the environments in which scientists and engineers work. While a scientist may be embedded in an organisation, such as a university or an industrial research department, the research activity itself, and its outcome, are determined by the scientist, and the outcome is published under his or her name. Engineers are embedded in industry, where they work in multidisciplinary teams, and where the result is usually attributed to the organisation; the engineer remains anonymous and invisible to society. The work of the scientist is judged by what new knowledge it produces; the work of the engineer is judged by how beneficial and useful it is. And if the scientific result later turns out to be incorrect; well, that is just further knowledge. Whereas if the engineer's work is incorrect, there is immediately a question of blame and liability. These, and many other differences between science and engineering make it clear that there must be fundamental differences between the philosophy of science and that of engineering, to the extent that the latter is even sensible.

Which brings us back to the confusion about the meaning of "technology" and its relation to science, and to the difference between technology and applications of technology. Technology, as the knowledge and the means of creating solutions to society's needs, is very much the realm and outcome of engineering. But what gets produced and is experienced by society are applications of technology, and they are essentially determined by the profit industry expects to be able to realise, which is the difference between the cost of production and the price the market is willing to pay; both of which are determined largely by social factors.

However, in a capitulation to common usage, we shall continue to use such expressions as "the influence of technology", even though it is the applications of technology that have an influence. Technology, just like any form of knowledge, has no influence, and its value is like that of money; it is only realised if and when it is used.

The involvement and importance of technology in all aspects of society, ranging from the most mundane items of daily life, such as brushing teeth or going to the toilet, to international relations and power struggles, as exemplified by the China-US competition, is so obvious and ubiquitous that we no longer think of technology and its applications as something separate from us as humans. But if you close your eyes to shut out all the objects surrounding you, and just look inward at yourself, there is a being with the ability to sense its environment, to think, to remember, to form opinions and make judgements, to create objects (*homo faber*), and, above all, with the ability to relate to and communicate with other humans, and thereby form societies. In principle, all of these abilities exist completely independently of any technology; but we have become so impressed by, and dependent on the results of our ability to create things that we have lost sight of the primary purpose of creating them: To support and enhance our abilities as human beings, and that the greatest of these abilities is not the ability to create things, but the ability to communicate and form societies. The progress of humanity has not been primarily in developing and exploiting technology, but in forming societies of increasing complexity and ability.

In *The Social Bond* (Sect. 2.3) there is a concise review of some of the literature relevant to the influence of technology on society (and vice versa), as well as some additional references. Here we are particularly interested in the relationship between technology and the stability of the evolution of society, and as an extension of the two cycles we considered in Sect. 3.5, we can see that there are two aspects to this relationship: The relationship between technology and society's belief system, and the relationship between technology and our ability to put changes of the belief system that we identified as fluctuations into effect. The latter is dominated by weapons technology (or, as it is often euphemistically called—defence technology), and as it is a subject matter of which I have very little knowledge, it will not be considered in this essay, except to note that an industry with an annual revenue exceeding 400 billion US dollars (https://sipri.com/databases/armsindustry) will be looking to create applications for its products.

Already in the creation of early settlements there was a level of technology and the ability to apply this technology; extracting and processing the materials required for tools and structures, the knowledge required to use the tools and to design and construct walls, roofs, water supplies, irrigation and drainage canals, streets, and clearing land for agriculture. Similarly, the enabler of the increase in both the flow and the storage of information has always been technology and its applications. For the interaction between settlements, the means of transport, in the form of horses, carriages, roads, and boats, would have been determining factors; again, reflecting a level of technology and its applications. With the advent of writing, somewhere around 5000 years ago, technology was applied in the production of the substrate on which the writing was recorded and in the production of inks and pens, etc., marking the beginning of an accelerating increase in information transmission and an increase in the ratio of technology-mediated transmission versus word of mouth. Thus, it appears that the level of technology and its application is a main determinant of μ, and the same is true of μ_a, in that the development of more capable and precise instruments drove the increase of our knowledge about Nature. However, the case of μ_0, which is exemplified by the scientist researching some aspect of Nature, makes us realise that it is not enough to have good instruments and facilities; there has to be the background knowledge and training which enable the scientist to work efficiently, as well as the time to do the work. And the same is true, to some extent, of the transmission of information; effective communications requires both knowledge and practice, and if you are scratching around from morning till night just to exist, there is not much time for acquiring either. So, we have arrived at the conclusion that, to a first approximation (or high level of abstraction), the dynamics of Θ is determined by the extent of technology application, and that there are two distinct effects involved. One is the effect of technology on the processing (e.g., printing) and transmission of information; i.e., on the rate of information presented to the individual. The other is the effect of technology on the conditions that allow the individual to process the information; effectively the standard of living in general. The former is the subject of the next section; the latter will be discussed in Sect. 7.1.

5.2 Information Technology

The history of technology is well known, and we might think of a time-line in terms of some of its major milestones, such as fire, tools, boat-building, agriculture, the domestication of animals, the wheel, concrete, metal extraction and processing, water power, chemical processing, steam power, electricity, telegraph, telephone, the internal combustion engine, pharmaceutics, electronics, and the Internet. But if we now pursue Fletcher's insight regarding the role of the stress introduced by interaction on the evolution of society, and want to identify this as one of the mechanisms by which technology influences evolution, we realise that there are two completely different epochs: before and after the development of telecommunications. In the earlier epoch, the interaction was based on direct, in situ interaction (the determining factor in Fletcher's density-based interaction limit) or on physical movement; either the movement of the persons involved, or by the movement of the physical carriers of information, such as letters, newspapers, and books. Certainly technology played a significant role, from horse-riding and boat-building to mechanised transport as far as the movement was concerned, and with printing as far as the physical medium was concerned, but both physical distance and the weight of the medium were determining factors in the interaction and in its effect on the evolution of society. But once the telecommunications industry started its exponential growth this changed, and today the situation in the developed part of the world is characterised by the following features:

a. Individuals have instant and unlimited access to a global communications network carrying voice, video, and data.
b. The cost of transmission is largely independent of volume and distance (e.g., fixed price contract for unlimited volume).
c. The capital invested in the telecommunications infrastructure is a significant part of the total investment in fixed capital.

As a result, the strictly spatial dependence of the behavioural stress, as in Fletcher's I-limit, is diluted by a dependence on the information environment, characterised by the parameter μ in our model. In an extreme case, a person could avoid all direct contact with other people and interact only via the telecommunications infrastructure, in which case the dependence on population density would disappear.

While face-to-face interaction is still important, in particular in conveying sentiments, the flow of information impinging on the individual is increasingly mediated by technology. And it is useful to divide the information and the mediation associated with it into two groups. One is the one-to-one interaction, mediated through writing (letters and email), telegraphy, telephony, and video; the other is the one-to-many interaction, mediated through printing, radio, TV, and the Internet, and with the mobile phone, or tablet, as the iconic user device in both groups. In the first group, the role of technology is to make the transmission of information more convenient (mobile, small device), capable of a great variety of content (speech, pictures graphics, video), and inexpensive (certainly on a bit-kilometre basis). In the second group, in addition to this same role in transmitting information, technology has a role that is becoming by far the most important one, and that is in providing the ability to manipulate the information in real time. This manipulation takes on several well-known forms, including

- Storage and retrieval. Until recently, the main means of this was to store the information as printed matter in libraries and other repositories. Now the collection and recording of information by electronic means, and the proliferation of these relatively low-cost means, such as surveillance cameras, mobile phones, and Internet-enabled devices and equipment, combined with rapidly decreasing cost of storage, is resulting in data bases of various kinds; both publicly available, such as the data stored on web sites and accessible via search engines, as well as specialised ones, available by subscription or for a fee, for practically every conceivable subject. However, many access channels, such as Internet browsers, mix the retrieved information with advertising, and also attach an evaluation to the information by influencing the order in which items are presented.
- Editing. By reorganising the sequence of items making up a message, the meaning of the message can be changed. And in this manner it is also relatively easy to imply an attribution that is not correct. But it is important to realise that it is not necessary to modify or falsify information in order to change its meaning; equally effective is simply deleting or ignoring part of the information or certain subjects that contradict the desired message. This is very noticeable in the Australian media when it comes to subjects that reflect adversely on Western interests or behaviour. However, this approach is not without its risk of backfiring, as was noted by Robert Louis Stevenson: "The truth that is suppressed by friends is the readiest weapon of the enemy" [5].
- Commenting. By interspersing the factual information with the provider's comments, the impact, and even the meaning, of the information can be changed. In particular, it is often difficult to discern what is fact and want is opinion.
- Packaging. The processing and combination of information into consumable "stories" is more convenient than ever, as evidenced by a plethora of newsletters and blogs. The use of the network is moving from interpersonal communication to the provisioning and accessing of *content*, at the same time displacing printed matter, and the distinction between μ and μ_a is becoming less a matter of the type of information than of the individual's *intent*, as "being immersed in society" is becoming more like "being immersed in the web".

Some fairly fragmented data on the growth and current extent of IT, defined as all applications of technology that support the generation, acquisition, transmission, processing, and display of information, is provided for reference. To get a feel for the magnitude of information flow and of its rate of increase (average yearly rate shown in bold font), here are some values presented in [6]:

World's technological capacity to store information (in exabytes $= 10^{18}$ bytes):

1986 2.6   **25%**

1993 15.8

2000 54.5

2007 295.0

World's technological capacity to receive information through one-way broadcast (in exabytes):

1986 432   **7%**

1993 715

2000 1200

2007 1900

World's effective capacity to exchange information through two-way communication networks (in exabytes):

1986 0.281 **30%**

1993 0.471

2000 2.2

2007 65

The above data values assume optimally compressed data, but that does not need to concern us here. What is of interest is that the yearly increase, shown in bold for each group of data, in the broadcasting of information, as in radio and TV, is much less than for interactive communications; a reflection of the decreasing cost of transmission. And monthly internet traffic, which grew from 1 exabyte in 2004 to 21 exabytes in 2010 (an estimate provided by Cisco), or a yearly increase of about 66%, is an indication of the importance of being able to access content. But irrespective of the accuracy of these estimates and of their detailed interpretation, they demonstrate the fact that the flow and availability of information, represented by μ and μ_a in our model, are increasing exponentially.

A different view of the application of IT emerges if we consider it from a business point of view, as the IT industry, and characterise it in economic terms. A problem with this is that the term "IT industry" is not very sharply defined, but we might start with a research report provided by CompTIA (https://comptia.org, drawing on various sources), which views the IT industry as consisting of the following components:

IT Hardware (29%)

- Servers
- Personal computers
- Storage
- Smartphones
- Tablets
- Network equipment
- Printers and other peripherals.

IT Services (18%)

- Planning and implementation
- Support services
- Operations management
- Training.

Software (12%)

- Applications
- System infrastructure software.

Telecom Services (41%)

- Fixed voice
- Fixed data
- Mobile voice
- Mobile data.

The percentages given for each major component are out of a total global revenue of US$3.8 trillion in 2016, which could be compared with a global GDP of US$78.6 trillion. However, it is important to be clear about what the 3.8 trillion refers to, as the following figure, Fig. 5.1, demonstrates.

In Fig. 5.1, the block "IT-based services" is by far the most significant one, and it is composed of a number of different types of organisations. One is the online retailer, of which Amazon is the most prominent representative, with a revenue of 100 billion and a capitalisation of 482 billion US$. The company sells directly from its own warehouses, but does also provide an online marketplace for some other retailers. The online marketplace is the second type of organisation, of which Alibaba (capitalisation of 455 billion US$) and eBay are prominent representatives; these companies act as middlemen between buyers and sellers, and get their revenue mainly from advertising on their various web sites. A third type are search engines, such as Google (revenue 70 billion and a capitalisation of 600 billion US$), that provides a free service to its users, paid for by advertising. The fourth type are the various social media, such as Facebook (revenue 35 billion and a capitalisation of US$500 billion), which get their revenue from advertising. A fifth type consists of a wide range of specialised information providers, often in the form of newsletters or reports, but also as online versions of journals and newspapers; partly free, partly by subscription or purchase, and mostly with additional revenue from advertising.

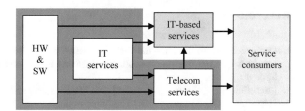

Fig. 5.1 The entities involved in IT-related activities

There are, of course, a whole host of companies providing specialised online services, such as email marketing (4.5 billion US$ yearly revenue, according to Transparency Market Research), accounting, tax returns, and the like. And, finally, there are a number of companies providing online payment services, obtaining their revenue from a percentage of the payment amounts.

The three blocks in the dark area of Fig. 5.1 are the ones accounting for the 3.8 trillion US$, and out of this sum the 41%, or US$1.56 trillion, of the hardware and software block presumably represents the gross fixed capital formation of the IT infrastructure of the telecom and IT-based services industry.

The three main points to take away from this are that, through their business activities, these companies provide a vast amount of information, a considerable part of this information is in the form of advertising (that is, its purpose is to persuade), and their combined capitalisation makes them a significant part of the economy.

A final indication of the present role of IT-mediated information exchange in our society is the amount of time we spend on IT-based activities, including fixed and mobile voice and data, TV, and radio. There are various estimates, e.g., from eMarketer (www.emarketer.com/) and Hacker Noon (www.emarketer.com/), as already noted in Sect. 3.3, and while they are fairly poorly defined, it would appear that in the US people spend several hours per day watching TV and as much as 4 h per day on mobile devices, of which 1.5 h on social media and 2.5 h on apps. This is certainly considerably more than they spend on face-to-face interaction.

In summary:

a. For the average person in the developed world, information via some form of technology-supported medium has become the dominant form of information input—the dominant component of μ in our model—and the time spent in interacting with these media is taking up a significant and increasing part of the time available to us for processing information.

b. Advertising is a large and increasing part of the information input.

c. The ability of advertising, and other forms of information aimed at persuading the individual, to catch the attention of the individual is increasing due to personalised targeting, with the data required for this targeting mined from the individual's activities with electronic media.

d. A further effect of the personalised targeting is the increased persuasiveness of the messages.

References

1. Dessauer F (1956) Streit um die Technik. Josef Knecht, Eisenbach
2. Mitcham C, Li B, Newbery B, Zhang B (2018) Philosophy of engineering, east and west. Springer, Berlin
3. Michelfelder DP, Newberry B, Zhu Q (2017) Philosophy and engineering. Springer, Berlin
4. Fritzsche A, Oks SJ (eds) (2018) The future of engineering. Springer, Cham

5. Stevenson RL (1890) An open letter to the Reverend Dr. Hyde of Honolulu. In: Lay Morals. Published by Chatto and Windus in 1911. Available online from Adelaide University at https://ebooks.adelaide.edu.au/s/stevenson/robert_louis/s848lm/
6. Hilbert M, López P (2011) The world's technological capacity to store, communicate, and compute information. Science 332(6025):60–65

Chapter 6
The Role of Education

Abstract Education is the other main component of the environment in which the evolution of society takes place; it provides the essential transmission of information from one generation to the next. The role of education is explored, starting with Durkheim's view, but then progressing into the importance of critical thinking, before investigating how education—as a process—influences the evolution of society. Finally, the relationships of education to technology and to economics are acknowledged.

6.1 A Brief Review

The importance of education in supporting both the maintenance and further development of society has been recognised for a long time—at least since the ancient Athenian society. But the understanding of what the word "education" means, including such aspects as its purpose, how it realises this purpose, the degree of its importance, who is involved in it, and what activities it comprises, has varied greatly over time and in different societies, as exemplified by Athens and Sparta. For our purpose it will prove useful to characterise education according to its purpose and the framework in which it is delivered, resulting in four main types, as shown in Fig. 6.1.

In the classification of purpose, "Social" signifies that the purpose of the education is for the benefit of society; for the recipient to become a productive citizen, in the sense of providing an economic return, as well as in the sense of providing an input to the stability and orderly operation of society. "Personal" signifies that the purpose is to develop the individual's sense of uniqueness and that the nature of the individual's relationship to society is more than just as a cog in the machinery;

Framework	Purpose	
	Social	Personal
Institutionalised	A	B
Open	C	D

Fig. 6.1 The four types of education, characterised by the purpose of the education and by the framework within which it is delivered

© The Editor(s) (if applicable) and The Author(s), under exclusive license to Springer Nature Switzerland AG 2020

E. W. Aslaksen, *The Stability of Society*, Lecture Notes in Networks and Systems 113, https://doi.org/10.1007/978-3-030-40226-6_6

77

it should be a conversation with society. Thus, the purpose could be expressed as developing the individual into a valued partner in this conversation, and it should enable self-fulfilment and the enrichment of the person, allowing the individual to appreciate and contribute to the culture in which it is embedded, without necessarily having any direct economic effect. In the classification of framework, "Institutionalised" indicates that the education is provided within a framework institutionalised by society and with attainment resulting in a formal degree or qualification recognised by society, and "Open" encompasses all other forms of information input with the potential to influence a person's behaviour.

The two "coordinates"—framework and purpose—were chosen here because the framework can be related to how technology mediates the education process, and the purpose is a determining factor in how education influences the evolution of society. This choice is by no means unique; in particular, it completely ignores the whole issue of *how* education is provided, i.e., pedagogy. Nor are the two coordinates completely orthogonal, but they will provide a useful space in which to analyse the relationship between education and the evolution of society. A different classification is one according to the content of what is being transmitted in the education process. At the basic level it is *facts*—anything from the names of things to the values of their parameters. At the middle level, it is knowledge about how to *apply* the facts, e.g. how to design or create something—from a cake to a space station. At the top level, it is knowledge about how to *evaluate* the facts; *critical thinking*, to which we shall return shortly.

To relate this high-level taxonomy of education to the existing body of knowledge on education and its place in society, a good place to start is with the overview provided by four essays by Emile Durkheim, published in the early part of the twentieth century, and published in English by The Free Press under the title *Education and Sociology* [1]. As a sociologist, Durkheim approached education as a characteristic of society, the means by which it secures its own existence, and as a result, each society has its own form of education. However, while he recognised that education could be extended throughout a person's life by social interaction, his focus was more narrow, and he defined education thus:

> Education is the influence exercised by adult generations on those that are not ready for social life. Its object is to arouse and to develop in the child a certain number of physical, intellectual and moral states which are demanded of him by both the political society as a whole and the special milieu for which he is specifically destined. (ibid. 71)

Durkheim elaborates on this by saying that we can differentiate between two sources of our behaviour: one is made up of the mental states that apply only to ourselves and to the events of our personal lives, the other is made up of the ideas and practices that reflect the society in which we exist, such as religious and moral beliefs, and generally collective opinions of every kind. It is these latter items that form the social being, the development of which is the goal of education; "It is society that draws for us the portrait of the kind of man we should be, and in this portrait all the peculiarities of its organization come to be reflected." (ibid. 123).

This emphasis on education reflecting the needs of the particular society, which he illustrates with the development from primitive, tribal societies, through Greek, Roman, Middle Age, and Renaissance societies, up to modern society, is compelling. But he does not enter into why society developed, what drove this development, and when he describes "the egotistical and asocial being that has just been born", he seems to miss the fact that it is this being that has created our present society. In the view expressed in our model, besides its physiological features and capabilities, this being is born with an information processing capability that, generally speaking, is capable of evaluating externally supplied information based on stored knowledge and a single criterion: survival. What constitutes survival, and the strategies for achieving it, is developed through interaction with the individual's environment and, by our unique ability to communicate, this knowledge has been transmitted from generation to generation, continually enriching that environment until arriving at today's complex society, in which survival is not only survival of the individual, but increasingly the survival of society itself.

However, Durkheim then goes on to effectively reduce the importance of the this initial nature by stating "Now, if one leaves aside the vague and indefinite tendencies which can be attributed to heredity, the child, on entering life, brings to it only his nature as an individual. Society finds itself, with each new generation, faced with a *tabula rasa*, very nearly, on which it must build anew." (ibid. 72). And he comes back to this point in several places; e.g. "But fortunately one of the characteristics of man is that the innate predispositions in him are very general and very vague." (ibid. 82), and "But only very general, very vague dispositions, expressing the characteristics common to all individual experiences, can survive and pass from one generation to another." (ibid. 84). With these qualifications, his view essentially conforms to the fourth of our simplifications, and affirms the pre-eminence of education over heredity.

With regard to the types of education defined in Fig. 6.1, Durkheim's concept of education is definitely as an institutional activity, provided by professional teachers. However, it is when it comes to the purpose of education that the relevance and usefulness of his analysis to our model requires an amount of interpretation. He recognises two components of education—one with general content applicable to all, and one with specialised content, subdivided according to professions, or to "the special milieu for which he is specifically destined"—but the purpose of both is social, in the sense of providing a dividend to society, and together they create the social being. The specialised component is clearly in the social column; i.e., of type A, but the purpose of the general component is open to some interpretation; in particular, given the context in which it was provided (France around 1900). Durkheim states that the purpose of the general content is to make a person conform to the sentiments and institutions of the particular society; to make the person a good citizen. "Education is, then, only the means by which society prepares, within the children, the essential conditions of its very existence." (ibid. 71). This places the general component also in type A, but if we reformulate the purpose as "make the person a valuable contributor to the development of society", the component would be classified as type B. With this, we could interpret Durkheim's general component of education as concerned with the strategy for survival and, as the information processed by the two components of education corresponds roughly to our classes 2 and 3, we can say that the purpose of the general component is to create the individual's identity. Throughout this collection of essays, Durkheim focuses on the importance of the second component, and it is this that makes his observations so relevant to our model.

The first thing to note is the mutual interaction between society and the individual, with education as the most important mediator:

> Whereas we showed society fashioning individuals according to its needs, it could seem, from this fact, that the individuals were submitting to an insupportable tyranny. But in reality they are themselves interested in this submission; for the new being that collective influence, through education, thus builds up in each of us, represents what is best in us. Man is man, in fact, only because he lives in society. (ibid. 76)

> Thus the antagonism that has too often been admitted between society and individual corresponds to nothing in the facts. Indeed, far from these two being in opposition and being able to develop only at the expense of the other, they imply each other. The individual, in willing society, wills himself. (ibid. 78)

Except for the phrase "what is best in us", which seems to be in conflict with the *tabula rasa* view of the newborn child, this expresses the idea of the ongoing interaction between individual and society. The role of education is not to make the individual conform to a fixed set of rules, but to ensure that the accumulated wisdom provides the foundation for each individual to contribute to society as it evolves from generation to generation This evolution is not progressing toward some externally given ideal state; it is a progress without limit, We could therefore interpret "what is best in us" as the ability to discern what is most likely to support the survival of society; i.e., the collective intelligence.

Durkheim emphasized a historical approach to understanding education, recognising the changes that have taken place in societies over time, and the many distinct societies that have existed, and how each one placed different requirements on education. In particular, on how to ensure that the individual developed into a social being that was able to contribute to the maintenance of a stable society. But there is much less, if any, emphasis on what characteristics the individual should have in order to be able to contribute to the *dynamics* of society. Society does not present the individual with a static environment; the individual is continually exposed to fluctuations, which can be of economic, political, and social nature, and the manner in which it responds to the fluctuations determines the evolution of society. This issue is taken up by Richard S. Peters in the essay *What is an educational process* [2], where he says "… activities like history, literary appreciation, and philosophy, unlike Bingo and billiards, involve forms of thought and awareness that can and should spill over into things that go on outside, and transform them.", and "All forms of thought and awareness have their own internal standards of appraisal. To be on the inside of them is to both understand this and to care. Indeed the understanding is difficult to distinguish from the caring; for without such care the activities lose their point." (ibid. 78), and "(For an educated man) his knowledge and understanding must not be inert either in the sense that they make no difference to his general view of the world, his actions within it and reactions to it *or* in the sense that they involve no concern for the standards immanent in forms of thought and awareness, as well as the ability to attain them." (ibid. 79).

Peters then proceeds to develop this issue in term of *principles*. "Understanding principles does not depend on the accumulation of extra items of knowledge. Rather it requires reflection on what we already know, so that a principle can be found to illuminate facts. This often involves the postulation of what is unobservable to explain what is observed. So it could never be lighted upon by 'experience'." (ibid. 18). Principles are close to what we have identified as the information items in the identity, and the salient concept here is *reflection*. We introduced this in connection with our model, in Sect. 3.6, and it is central to the evolution of society. It is the process that allows us to make sense of information about changes to society and to take adaptive action, where required. The process of reflection has two sides to it; one, establishing relationships between information items and forming principles, and, two, assessing the applicability and correctness of existing principles and assumptions, both as they stand and in the light of new information items. With regard to this latter activity, *critical thinking*, Peters makes two observations. The first is that while it is desirable to acquire a critical attitude, critical thinking is no substitute for action; only the two together lead to effective improvement. The second is that critical clarification of principles is a very different exercise from applying them in concrete circumstances, which requires judgement, something related to a critical attitude, but developed mainly through experience, rather than through a thought process about principles. These issues, as they relate to education, are also treated in the last chapter in the same book, *On teaching to be critical* [3].

A more recent book, *Educating Humanity:* Bildung *in Postmodernity* [4], treats reflection and critical thinking under the perspective of the concept of *Bildung*, a product of the neo-humanist tradition that flourished in Germany between 1770 and 1830, and expressed the ideals of the Enlightenment. It is related to the concept of an educated person, as contrasted with a trained person, but *Bildung* has two sides to it. One is the content, in the sense of knowledge—what constitutes "a good education"—as in a list of what one should have read and studied; the other is a capacity for discerning the reality behind what presents itself as necessary, natural, general, and universal, an understanding of how the world operates. As such, *Bildung* as a process was very much one of self-*Bildung*, one of a personal quest to achieve rational autonomy, or, in terms of our model, one of developing one's identity. It could be seen as developing a capacity for critical thinking, except for the fact that the concept of *Bildung* was based on the assumption that there were concepts and principles that were general or universal, and also enduring, and the assessment of the world was tied to this measure. As this assumption came under increasing scrutiny in the 19th and 20th centuries and the interest shifted to more down-to-earth behaviouristic concepts of capabilities, the concept of *Bildung* lost much of its relevance, but the authors of this book examine to what extent and in what form it can be resurrected, given that it contains much of value.

Relaxing the requirement for the core concepts and principles of Bildung to be enduring, we can still ask if there are, at each point in time, concepts and principles that are characteristic of society, in the sense of forming a basis for the evaluation of everyday experience. This is the question addressed by Gert Biesta in Chap. 4, *How General can* Bildung *Be? Reflections on the Future of a Modern Educational Ideal* [5], and he approaches it by considering two different answers. The first is the *epistemological* answer, in which the general is understood as universal; the second is from the *sociology of knowledge*, in which the general is understood as a social construction. In the first, knowledge is considered to be general if it is valid and can be applied everywhere and at any time; it is an intrinsic quality of knowledge and is taken as proof that it represents reality 'as it is'. The existence of such knowledge is demonstrated by the omnipresent success of technology based on science; what Biestra calls *the technology argument*. In the second, the general is only an expression and product of social relationships, and hence an expression and reinforcement of the way in which power is distributed at a certain moment in time, so that the general is in fact a manifestation of the particular at a specific time.

Biestra provides valuable insight into problems associated with both of these answers, as well as with an extension of the second one—the *critical theory* of *Bildung*. But, in view of these problems, he then goes on to propose a different understanding of the general, based on work by Bruno Latour. As with some of the other work of philosophers of technology, this work of Latour's appears highly artificial to an engineer, and reflects a poor understanding of the relationships between science, technology, engineering, and industry (see also Chap. 5). In the context of our model, sociology of knowledge (all knowledge in class 3 is a sociological product) together with critical theory, as described by Biestra, provide an adequate definition of the process side of *Bildung*. The involvement of critical theory is further elaborated in Chap. 5, Bildung *and Critical Theory in the Face of Postmodern Education* [6]. "This article attempts to show that the prospects for today's resistance to the current process of dehumanisation, which flows from normalising education, are closely bound up with the project of *Bildung* and its re-articulation by the critical thinkers of the Frankfurt School." (p. 75). A central point in this re-articulation is the autonomy of inwardness of the individual and an emphasis on the importance of reflection as a precondition for, and in a certain sense the first manifestation of, the resistance to the hegemonic order of things. Transcendence here takes on almost the opposite to its religious meaning; it is the possibility of overcoming the normalisation and hegemonic order.

Chapter 6, *On Irritation and Transformation: A-telelogical* Bildung *and its Significance for the Democratic Form of Living* [7], contains further insight into the meaning and relevance of *Bildung* in today's society Reichenbach starts out by saying "The question of the meaning of the (old) idea of *Bildung*—as the refinement of intellect, sensibility and judgement—in a postmodern society can be approached by starting with a characterisation of postmodernism as a tired, even exhausted modernity; a modernity that has lost faith in its own dreams and promises (although it may nevertheless be able to evaluate these positively)." (p. 93). This may be a somewhat pessimistic view; the fact that the "dreams and promises", in the sense of a blissful end-state, has been abandoned, does not mean that the process part of *Bildung* has lost its relevance and importance. And Reichenbach acknowledges this when he asks

"Could it be that there is only one plausible goal left to postulate by a late-modern theory of *Bildung*? I am referring to the individual's ability to initiate processes of self-transformation." (p. 95). That he sees self-transformation as a process without any hope of success, in the sense of not reaching an end-state, is really irrelevant; it is one of many processes whose success is measured by their ongoing productivity rather than by any change. Indeed, most, if not all, of the processes, both physical and mental, in the human body are of this type. The productivity of self-transformation can be seen as the opposite to acting through habit and "everyday thoughtlessness", a term he attributes to Eugen Fink (p. 100).

Reichenbach introduces the concept of *self-deception*, and stresses the importance of accepting this possibility in two directions. In the inward direction, the individual needs to accept that, despite its best and sincere efforts, the view of life arrived at may be subject to valid objections; otherwise *Bildung* degenerates into dogmatism. In the outward direction, every individual, and thereby society as a whole, needs to accept and make allowance for self-deception and see it as an expression of society's vibrancy rather than as something resulting in unnecessary or disproportionally severe social sanctions that only serve to cause the self to harden (p. 100). This dual acceptance becomes a prerequisite for the proper functioning of what we have called the averaging process within the operation of the collective intelligence, and thereby for the democratic form of living.

As a conclusion to this very brief and selective foray into the sociology and philosophy of education, we find that the process part of *Bildung*, in its postmodern form, is the component of education that has the greatest impact on the evolution of society; it is the component that constitutes the major part of type B (and, as we shall see, type D) education, and that enables the process of reflection. So far, we have, more or less by implication, considered education in an institutional framework. However, the fact that the individual's interaction with society throughout its lifetime includes elements of education, and that, as a consequence, the identity keeps evolving, has always been the case. Indeed, before institutionalised education became significant, it was the main form of education. It was always recognised as a factor in *Bildung*, and from the introduction of printing it has gained in importance due to the enabling effect of technology, as we shall consider in Sect. 6.4.

6.2 Education as a Process

6.2.1 Lifecycle of the Individual

In order to be able to focus on the role of education in the evolution of society, which, as a transfer process from society to the individual, involves the dynamics of the information processing within each person, we need to distinguish between a macroscopic (social) view and a microscopic (individual) view. So far, our model has not explicitly distinguished between the two views, and the parameter α was a

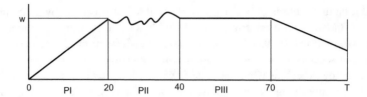

Fig. 6.2 The development of the identity over the individual's life time. PI is the period of schooling, PII is the period of adult development, and PIII is the period of intergenerational interaction. The time-scale is indicative only

characteristic of both the individual and the social identity; now we have to distinguish between what we shall call the *collective identity* and the identity of the individual, with the former being the average of the latter. Even if society is unchanging, each individual progresses through its lifecycle, and to look at the information processing that accompanies this progress at the individual level, we need to identify individuals by their age. Denoting the life expectancy of persons in the society by T, we can, as far as education is concerned, distinguish three periods within this lifetime, as shown in Fig. 6.2.

The duration assigned to each of the periods in Fig. 6.2 should be understood as approximate and representative of the life cycle in a developed society. The significance of each period is as follows, and the assumptions made in these descriptions constitute further simplifications of our model of society:

PI is the period in which the individual's identity is initially formed through the acceptance of information received from authoritative sources. First from parents, and then through a formal, institutionalised education, i.e. schooling; both based on the existing state of the collective identity. The size of the identity rises to its full value, w; the value of α will depend on both the socio-economic structure of the society and on the schooling system and the content of its curriculum, but would normally be high throughout the phase.

During PII and PIII, the individual is exposed to interaction with other individuals, with a resultant value of the level of conflict in the identity, β, but is also engaged in making its own observations and evaluations. This is the wider scope of education; the individual develops a personal character and, in doing so, creates new information items, with the result that the value of α decreases. The individual provides the input to the schooling and development of the next generation, transferring both old and new information items developed during these phases, and thereby contributing to the evolution of the collective identity. However, in these phases the individual performs another, very important task; evaluating input received from other individuals and commenting negatively on those items that conflict with the individual's identity (which, of course, contains, in its α-part, the items common to all individuals). The effect of this, together with the averaging performed to create the collective identity, is to dampen any sudden, large changes to the collective identity. The change to the individual's identity is less in PIII than in PII; this is due to (at least) three factors: The attention factor, x, introduced earlier, becomes smaller with age; a matter of *déjà vu*. The persuasiveness of new information items, our parameter p, is also reduced with age; we become more sceptical. And, finally, our personal questioning and investigating activity, our parameter μ_0, decreases; we become more resigned. As a consequence, and with reference to Sect. 3.6, we shall assume that α remains constant in PIII, and that the value of β is small.

The use of the term "generation" reflects its use in the vernacular; in the process we have developed, the evolution of society is continuous. The perception of generations in the evolution of society arises rather from major events, such as world wars and the introduction of disruptive technologies, e.g. steam power, telecommunications, and computing. However, it is obvious that the process of education introduces a time constant into the evolution of society, and we can easily discern two contributors. One is the duration of PI; as the extent of our knowledge increases, the time needed for reaching a level of competency that will allow the individual to perform satisfactorily as a member of society has been increasing, despite greater specialisation.

The other contributor is the duration of PII; a period in which the information acquired during PI is put to the test and modified by experience and new information arising from the interaction with other individuals. This process decays with time from the end of PI as the identity "hardens"; at some point becoming negligible and thus marking the end of PII. But the decay is counteracted by the increase in leisure time, combined with the improved access to information through information technology, with the result that there has been an increase in the duration of PII. It is interesting to consider that as the duration of the phases increased, due to both increasing life expectancy and complexity of society, the relative duration of the three phases remained largely unchanged, which would be interpreted to mean that society needs a balance between the functions performed by people in the different phases in order to stay functional.

6.2.2 Inter- and Intra-generational Information Transfer

In the essays we referenced in Sect. 6.1, Emile Durkheim emphasized the nature of institutionalised education as a transfer of knowledge from one generation to the following one, and while this is being somewhat blurred by the decrease in teacher/student interaction in schooling, the availability of information for ongoing non-institutionalised education, and the educational role of social media, the delay inherent in this transfer is still an important factor in the dynamics of society. To get a feel for the effects of this delay, we can construct a highly simplified model of the information transfer, based on the individual lifecycle defined in Fig. 6.2 and, as we are only considering information in class 3, on the concepts of the individual identity and the social cohesion, α, as introduced in Sect. 3.4. In Fig. 6.2 we gave some indicative durations of the three phases; for simplicity, we shall now assume that the three phases have the same duration, L. This allows us to divide time into segments of duration L. During one segment, the institutionalised education transfers the content of the α-part of the identity of the individuals in the phases PII and PIII

T	U(T)	V(T)	W(T)
0	X	X	X
1	$rX+D_1(1)$	$rX+D_2(1)$	$rX+D_3(1)$
2	$r(rX+D_2(1)+D_3(1))/(2-r)$ $+D_1(2)$	$r(rX+D_1(1))+D_2(2)$	$r(rX+D_2(1))+D_3(2)$
3	$r(r^2X+r(D_1(1)+D_2(1))+D_2(2)$ $+D_3(2))/(2-r^2)+D_1(3)$	$r^2(r^2X+r(D_2(1)+D_3(1)))/(2-r)$ $+rD_1(2)+D_2(3)$	$r^2(rX+D_1(1))+rD_2(2)+D_3(3)$

Fig. 6.3 The evolution of the α-part of the identity from an initial content of X under the influence of the phase-internal processes of reflection and discovery

(society's belief system) into the identity of the individuals in phase PI. Let U(T) be the information received by individuals in PI, V(T) the information transferred from individuals in PII, and W(T) the information transferred from individuals in PIII. If there is no new information generated *within* a phase, then $V(T + 1) = U(T)$, and $W(T + 1) = V(T) = U(T - 1)$; that is, education only preserves the *status quo*, and society does not evolve. But in Sect. 3.2 we introduced the activities that result in the information flow we labelled as *discovery* in Fig. 3.1. Let us denote the information generated from these activities during the time segment n in phase i by $D_i(n)$, with $i = 1, 2, 3$, but assume that the *amount* of this information is the same in each phase and time segment, and denote it by D. Furthermore, as the size of identity, w, is constant (as we explained in Sect. 3.3), and, if the evolution is in a stable state, α is also constant, it follows that the amount D *replaces* a fraction of the information from the previous phase, reducing that amount of information by a factor r. Now, let the information in each phase in the initial time segment be identical and its amount be X, so that $r \cdot X + D = X$, and let the phase-internal activities be switched on at the beginning of time segment 1; we then get the temporal development shown in Fig. 6.3.

The temporal development displayed in Fig. 6.3 can be characterised by how the original content of the identity, i.e., of society's beliefs, in the material that is propagated by education decays over time; this is the proportion of X in U(T). If we denote that proportion by R(T), Fig. 6.3 allows us to deduce that

$$R(T) = \frac{r^T}{2 - r^{(T-1)}}. \tag{6.1}$$

We can then define the half-life of society's belief system as the value of T that makes $R(T) = 0.5$, and this is shown in Fig. 6.4 as a function of the intensity of the generation of new information, $1-r$, measured as a fraction of the existing information, for a choice of the unit of T (i.e. the duration of a phase) of 20 years.

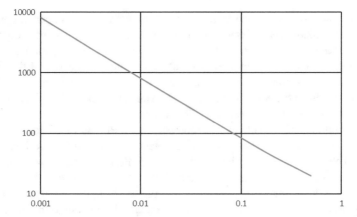

Fig. 6.4 The half-life of the α-part of the identity—society's belief system—in years, as a function of the intensity of the generation of new information during a phase through reflection and discovery, measured as a fraction of the existing information. The duration of a phase, L, has been chosen as 20 years

6.3 Education and the Model of Society

In order to relate the evolution of society's belief system through the institutional, or schooling, part of education depicted in the previous section to the highly simplified model of society and its evolution we developed in Sect. 3.6, we need to consider two issues. The first one is that the ageing process, which transfers information from one generation (value of T) to the next, is a continuous process, but in Fig. 6.3 it is shown as a batch process. For example, in the cell V(2), the information rX + $D_1(1)$ is transferred from U(1) at the beginning of the time period T = 2. Then, during that period, the discovery and reflection activities that are performed by the individuals in phase PII of their lives result in new information, and the exchange and evaluation of this new information between the individuals during the phase results in the creation of new information in the α-part of the identity, $D_2(2)$, which replaces the same amount of information in the information present at the beginning of the phase. Thus, this updated information, which becomes part of the basis for education in U(3), has already undergone the evaluation and averaging process we identified as the operation of the collective intelligence, and it is this process, which we can consider to be society's immune system, which has a stabilising influence on the evolution of society. Any new idea, or item of information, even if meeting with approval in a local environment, needs to win approval of the wider society before it enters into the curriculum in PI; the "price" of this process is the delay it introduces, in the order of L. The rate at which the averaging takes place depends on the flow of information, μ_1, and on the persuasiveness, p. But our model, as developed in Sect. 3.6, is concerned only with the steady state, so that the value of p is what is needed to maintain a constant value of α in the face of a given flow of new information, μ_0, as we shall see shortly.

The second issue is that, while the education process deals with persons and their lifecycles, the model considers society to be composed of identical individuals, with the value of a parameter of an individual to be the average of that parameter over all adults in the real society, which allowed us to assume that the size of the identity, w, stays essentially constant. This means that, if we want to relate the model to the three-phase view of the education process presented above, we have to divide society into two sub-societies. The one, to which our model can be applied as it stands, is a society comprised of individuals based on persons in PII and PIII, and this is where the evolution of society takes place; where new information is absorbed into the identity. The other sub-society, based on persons in PI, is dedicated to education, and while the model concepts still apply here, the model would have to be completely changed to account for the behaviour in this phase. Also, the amount of evolution taking place in PI, although assumed to be identical in volume to that in the other two phases, i.e., D in all three phases, only becomes effective once it transitions into PII.

This process taking place in the PII/PIII society, in which α stays approximately constant, is what we consider to be the normal evolution of society. If $\alpha = 0$ there is no society, and if $\alpha = 1$ there is complete stagnation; essentially, a dead society. A healthy society would have a value of α somewhere around 0.5, and any significant fluctuation from this value would indicate a society in distress. If we now go back to Sect. 3.6 and Eqs. 3.5 and 3.7, and, as in Sect. 4.5 set $\nu = 1$, we have

$$\alpha\mu_0 = p\mu_1(1 - \alpha), \tag{6.2}$$

and then, with

$$D = L \cdot \alpha \cdot \mu_0 = L \cdot \mu_1 \cdot p(1 - \alpha) \tag{6.3}$$

and with $X = \alpha \cdot w$, we obtain

$$1 - r = \frac{D}{X} = \frac{L\mu_1}{w} \cdot p\frac{1 - \alpha}{\alpha}, \tag{6.4}$$

which is what we would expect. For a constant value of α, the product of the volume of new information produced in the duration of one phase, L, relative to the size of Θ, and the persuasiveness, p, determines the rate of change of the society's belief system.

Perhaps the most important aspect of these relationships becomes evident if we consider what we have been calling persuasiveness in more detail, and which we introduced in Sect. 3.6 as a defining characteristic of an information item. It is not a sharply defined concept, and while it is always related to the process of changing people's minds, it may be a measure of the effort that goes into it, but it may also be used as a measure of the efficiency of the process, or of its success rate, and so on. In a face-to-face situation, persuasiveness is comprised of many personal factors, such as tone of voice, facial expression, body language, and the use of humour; generally what is included in the concept of charisma. But as information transmission is increasingly mediated by numerous applications of technology, it is the means technology offers for increasing persuasiveness that becomes the dominant factor. There are numerous physical channels, both fixed and mobile, that allow the same message to be received by a significant proportion of society. And while the most recent developments are in the area of telecommunications, other means, such as printing (newspapers, flyers, billboards, etc.) have been around for some time. The format of messages is expanded from voice only to voice, images and videos, and music, as well as any combination of these, and most recently it is also possible to deliver the message via a two-way, or interactive, process. And the cost of repetition, which is a common means of increasing persuasiveness, is reduced. But perhaps the most significant aspect, as far as persuasiveness is concerned, is the ability to modify or edit the message in order to, on the one hand, make it resonate with characteristics of a particular audience, and, on the other hand, modify original content so as to change its meaning, e.g., by selecting only part of the content, or by rearranging the sequence of components of the content to reverse cause and effect. We can therefore ascribe a main part of the exponentially increasing rate of churn within society to the role of technology in increasing not only the flow of information, but also its persuasiveness.

There is another characteristic of the information flow (μ in the model) that is related to its persuasiveness, and that is its ability to capture the *attention* of the individual (the factor x in the model). Only a relatively small proportion of the information impinging on the individual is noticed and processed in any meaningful way, and it is a main purpose of advertising and any other form of thought-manipulation (including teaching) to increase this proportion; only then can persuasion commence. Here again, information technology plays a major role, often by the ability to combine different information items in the same message, e.g. creating a captive audience through apps that combine information with advertising, and so on.

Turning to the other sub-society—the one formed by the persons in PI of their lifecycles—the situation is quite different, in that rather than a small part of the information, D, being new, all of the information is new. And instead of the main process being a circulation of information, it is mainly a one-way path from teacher to student, steadily increasing the size of the identity, w. At any point in PI, α will have a particular value, and traditionally the aim has been to achieve a relatively large value, which, we now know, requires greater effort. However, the nature of the persuasiveness in this phase is somewhat different from what it is in PII/PIII. There persuasion meant changing an existing belief; in PI it means getting the individual to accept the importance of having a belief in the first place, and then to understand the importance of a shared belief structure in maintaining a viable society.

6.4   Education and Technology

As we did at the beginning of this chapter, we can start our consideration of the relationship between education and technology by introducing a structure to the relationship, and the first thing to note is that the relationship is a mutual one. Education influences technology (E → T), and applications of technology influence education. This latter influence has two distinct aspects: One is direct, in that Information Technology (IT) *enables* many forms of providing education (T → Ee). The other is indirect, in that technology *supports* the type of society and standard of living that allow people to engage in education (T → Es). Hence, we can identify twelve elements of the relationship, as shown in Fig. 6.5.

A brief description of each element is as follows:

a. All applications of technology require some form of knowledge and skill, and the earliest institutionalised form of transmitting this from one generation to the next would have been workshops, often organised into various forms of guilds. Today, the development and application of technology is completely dependent on a suitably educated workforce, from engineers to skilled tradespeople, and the dependence of technology on education is reflected in the current emphasis on STEM subjects.

b. Depending on the exact definition of the Personal purpose of education, we could say that this element includes the influence of ancient beliefs on technology, in the form of temples, pyramids, and other structures and artefacts expressing personal beliefs as propagated through religious education. But otherwise, the significant influence only arose in the last century or so, with institutional education starting to create a critical awareness and assessment of the role and effect of various technology applications until, today, attitudes to technology applications have become part of people's belief structure, and effect a corresponding influence on these applications (e.g. nuclear energy, GM foods, global warming, etc.).

c. One might think that this element would simply be a reduced version of element *a*, to the extent that Open education is (still?) a very much smaller component of education than the Institutional one, but this would be overlooking the fact that they are located in different periods of the individual's life. While the Institutional component is located predominantly in PI, the Open component is located predominantly in PII; that is, a period where experience is starting to reveal shortcomings in the state of technology. The effect of education in this period is therefore likely to be biased towards the development of new technology rather than towards the aims of element *a*.

Fig. 6.5 The twelve elements of the interaction between education and applications of technology

	A	B	C	D
E→T	a	b	c	d
T→Ee	e	f	g	h
T→Es	i	j	k	l

d. Much the same situation as in element *c* above, except that the subject matter is now not technology itself, but a critical examination of the influence of technology applications on society, and also that the experience and maturity of the individual is likely to amplify the effectiveness of this element.

e. This is the element most immediately associated with the concept of technology in education, and it has undergone a rapidly accelerating growth in the last decade. It offers the possibility of tailoring the education to a diverse student population (instead of e.g. having selective schools), of a more interactive delivery mode, and of flexibility of timing and location. An overview of these and other possibilities, as well as of some associated problems, are contained in [8], and some thoughts on the current enthusiasm are presented in [9]. Furthermore, the increase in the time spent on education noted in Sect. 2.4 is to a large extent the result of the demands of technology.

f. This is probably the element least influenced by technology; if anything, it is a negative influence. It is the element where the personality of the teacher has the greatest effect.

g. This element is very significantly influenced by technology; the Internet is presenting a plethora of sources of educational material that can be accessed outside of any institutional framework.

h. Besides a great amount of low-cost printed material for self-education and self-improvement, there are innumerable websites in the form of blogs, discussion groups, wikis, etc. that purport to be concerned with developing the individual's capability for critical thinking, but they are often promoting a particular point of view or degenerate into a free-ranging chatter with little, if any, educational value. It is difficult to say to what extent technology, beyond the level that printing reached fifty years ago, has had an influence on this element.

i. Institutionalised education depends on the ability of society to generate a surplus above what is needed for the existence of the individual, and this surplus is a consequence of the development and application of technology.

j. There is no in-principle difference between this and the preceding element of the relationship; in practice, the economic aspect of these two elements may tend to place more importance on the former.

k. In addition to the economic aspect (e.g. allowing people to pay for information and for IT services), there is the important role of technology in creating more free time (leisure time) in which educational activities can be pursued. The increase in leisure time was noted at the end of Chap. 2.

l. As for element j.

In the description of the interaction between education and technology above, it appears as a mutual interaction between two equal partners. But a moment's reflection shows that this cannot be quite true; education is a process of transmitting information, and that information must be created before it can be transmitted. That creation takes place within what we characterise as *industry*, in the widest interpretation of that concept; through scientific research flowing into the development and application of technology by engineers. Education then responds to a *demand* for the information created to be transmitted, both within the generation that created it as well as to following generations. The influence of education on technology is reflected in how well education responds to the demand, and that, in turn, depends to a certain extent on how well education utilises the means put at its disposal by technology. So, while the interaction between education and technology is complex, it is technology that is the driving force; this is just the current extension of Durkheim's analysis.

While the above is true in general, there is a difference in the relationship between the two frameworks in which education is delivered. Within the institutional framework, the institutions have a choice what, if any, technology they apply in providing the education, and they can, at least in principle, resist attempts by the technology providers to pervert the education process for their own commercial or ideological purposes. (The use of technology by an autocratic government to pervert the education process is a different matter.) Within the open framework it is much more difficult to differentiate between what is the provision of information for educational purposes and what is, essentially, advertising or propaganda, and the information technology offers more opportunities for presenting a message rather than factual information. Nevertheless, information technology is making the open framework increasingly important for providing education in phase PII, and within that offering are unprecedented means of focusing on the Personal purpose of education. A search for enlightenment and personal development that would, even as recently as a century ago, have been possible for only a small proportion of the population, and mostly in academia, is today available to anyone with Internet access. How many avail themselves of this opportunity is a different matter, but it is at least potentially a great force for change.

6.5   Education and Economics

Realising that there is a strong relationship between education and technology, and recognising how production is totally dependent on applications of technology, we would expect there to be a strong correlation between the level of education on the change in GDP *per capita*. This is indeed the case, and more for completeness than anything else, we review some of the available data.

We start out with a paper by E. A. Hanushek and L. Wössmann, *Education and Economic Growth* [10], mainly because they open their analysis with the observation that it is not only the attained level of education or the amount of resources spent on education that determine the impact on economic growth, but also the quality of the education, as in its effectiveness in producing cognitive skills. A straightforward correlation between years of schooling and GDP per capita growth suggests that each year of schooling is associated with long-run growth that is 0.58% points higher. But if the quality of the education is taken into account, which is now possible due to the increasing participation in standardised tests (e.g. Trends in International Mathematics and Science Study (TIMMS) and Programme for International Student Assessment (PISA)), it turns out that the variation in annual growth rates of real GDP per capita is explained mainly by the variation in education quality, with the effect of years of schooling greatly reduced.

Another measure is the education index, which measures a country's relative achievement in both adult literacy and combined primary, secondary, and tertiary gross enrolment, and it is clearly correlated with the per capita GDP, as demonstrated e.g. in [11], Fig. 5, which shows an exponential dependence of GDP on the education index. Further analysis in this paper shows that, for developed countries, the technology aspect of the education is dominant, and Habermeier concludes "The education in science therefore is a mandatory prerequisite for sustainable economic performance of a country." (ibid. 66).

An interesting approach to studying the relationship between education and economic development, and one that is based on the explicit involvement of technology in this relationship, is described in a paper by Yong Jin Kim and Akika Terada-Hagiwara, *A Survey of the Relationship between Education and Growth with Implications for Developing Asia* [12]. They focus on total factor productivity (TFP)—the part of output not explained by the amount of inputs used in production—as the most appropriate measure for analysing the influence of education on the economy. This is based on Sowlow's observation that long-run growth in income per capita in an economy with an aggregate neoclassical production function must be driven by growth in the TFP [13]. Denoting the TFP by A, the growth rate of the TFP is given by the expression

$$dA/dt = g_y - \alpha \cdot g_k - (1 - \alpha)g_h; \tag{6.5}$$

where

g_y growth rate of aggregate output;
g_k growth rate of aggregate capital;
g_h growth rate of aggregate labour; and
α capital share.

As the values of these quantities are all captured in the national accounts, the growth rate of the TFP can be determined for each nation. The results of the analysis are presented using two further concepts: the *technology gap*, which is the difference in the TFP of a nation and that of the most advanced nation (the TFP of the latter is called the *technology frontier*), and what we might call the *level of education*, as the ratio of tertiary education expenditure to primary and secondary education expenditure (the development in the paper goes via the relationship of education, or human capital, to technology, and is quite detailed). The main result is that, while TFP is positively correlated with expenditure on education, the influence depends on the relationship of the technology gap to the level of education. For a large gap, it is expenditure on primary and secondary education that is most effective, whereas for a small gap it is expenditure on tertiary education. But there is a limiting effect of the level of education, in that too high a level (implying too great a degree of specialisation) reduces the flexibility with regard to adopting new technology, and it is this insight that we shall use as an analogy in the next section.

6.6   The Challenge for Education

Emile Durkheim's definition of the purpose of education, which we met at the beginning of Sect. 6.1, is formulated in terms of society's requirements on the individual, something he considered to be both relatively fixed and well-defined in the France of 1900. Today, and in a society like the Australian one, that is no longer the case, and so the question of the purpose of education has been shifted into a question about the purpose of the individual. Durkheim's subdivision of the purpose into two parts, one general and one specialised, is still valid, but they both relate to the maintenance of (a static) society, and are contained in our types A and C in Fig. 6.1—the social component of the purpose. We have introduced a second component, the personal component, consisting of the types B and D in Fig. 6.1, additional to what Durkheim saw as the purpose of education. What we now have here is one component relating to society as a process, and one component relating to the dynamics of this process. Crudely put, the former component is driven by our desire to survive, the latter by our view of what should survive.

In Australia, there is currently a great deal of concern, both publicly and in the education community, about the declining proficiency of Australian students in the subjects grouped under the acronym STEM (Science, Technology, Engineering, and Mathematics), and there are basically two reasons for this concern. One is that the world in which we live is increasingly dominated by applications of technology, so that in order to understand that world and participate, as citizens, in deliberations about its management and development, a good understanding of these subjects is required. Without this understanding, the public discourse degenerates into a fruitless emotional, ideological, and acrimonial exchange. The other, more directly practical and measurable reason, is that the performance and competiveness of our economy is correlated with our performance in these subjects, as we saw in the previous section. Important as this concern is, it inevitably raises the more general issue of the purpose of education, and of the allocation of limited resources to the various subjects and structural elements that make up education. In particular, it raises the issue of the role of the individual in society, and of how education should prepare the new generation for this role. This aspect of education seems to become somewhat lost in the current debate, and it is this perception that is one motivation for the present essay.

This perception is by no means new or original, and it typically appears under the heading of sociology of education. A paper by Matthews [14], although focused mainly on some specific areas, such as gender, sexuality, ethnicity, and multiculturalism, gives a good overview of sociology of education in Australia up to 2010, and raises the various aspects of this perception. The historical trajectory goes from an initial focus on the practical problems of teaching, via increased interest in the relationship between education and social change, to a bureaucratic preoccupation with narrow utilitarian objectives and with testing and test results as the determinant of success.

Staying for the moment with Durkheim's subdivision of the social component, we realise that there has been a very significant change in the ratio of the information associated with the two parts of this component since his time. The information within the specialised part has been increasing at a rapidly accelerating rate, whereas the information belonging to the general part has changed hardly at all in volume. One popular measure (although of somewhat uncertain accuracy) of the accelerating rate is the time it takes for the amount of information to double; before Durkheim's time it was about 100 years; today it is estimated to be around 12 h. It is clear that this must have a profound effect, not only on education per se, but on our relationship to information in general, and the current situation is characterised by three main approaches. The first one is an increasing degree of specialisation; by narrowing "the special milieu for which he is specifically destined", the growth of the knowledge required to perform a profession is contained. But this specialisation has a downside, in that the communication between specialisations becomes more difficult and less effective; the interfaces between specialisations increase in number and present an increasing overhead. The second one is to increase the time spent on education, prior to entering the workforce, as documented in Sect. 2.4. However, as the individual's lifespan is limited, and the energy level declines at some point after the half-way mark, this increase can only have a very limited effect. And there is a further problem, in that the amount of information that can be usefully handled by the brain is limited; with too long an education period, what was learnt at the beginning has been forgotten at the end.

The third, and by far most promising way forward is to utilise the capabilities of the technology that is causing the information explosion in the first case, and to form a hybrid which combines human and artificial information processing capabilities in an optimal way. But there is another change looming for education and it is, in one sense, the core of what this essay is leading up to. It arises as an integral feature of the evolution of society and is, perhaps, best approached from the perspective of the individual's understanding of its role in society, which was introduced in Sect. 2.3. There, we proposed that a measure of this evolution could be the ratio of the individual's perception of its purpose as being the advancement of its own interests to it being the advancement of the interests of society, and while the former is inherent in our nature as individuals, the latte is acquired through education (in our wide interpretation of the concept). The problem we are faced with—the effects of which we can observe all around us—is that the rate of change of society, and the increase in its complexity, has led to a situation where the *actual* integration of the individual in society has outstripped our understanding (or acceptance?) of it. That is, the value of the above-mentioned ratio is higher than it should be for a harmonious operation of society, and this is where education comes in. In particular, the types of education we identified as Personal, i.e. B and D in Fig. 6.1. This may at first seem paradoxical: is it not the social types we need to promote? But this apparent paradox disappears once we recall the definition of the types. The social part of education. types A and C, has the purpose of preparing the individual to be a valuable contributor to society, to its maintenance and its growth. It is a view that sees the individual and society as two separate entities; society is something given, with certain well-defined characteristics, such as structures, institutions, and beliefs, that allow education to be fashioned so as to fit the educated individual into it. This is, essentially, Durkheim's view, and one that has maintained much of its influence until today. It is a situation very similar to thermodynamics, which, despite the "dynamics", relies on the change being so slow that it is a good approximation to assume that the system under consideration is in equilibrium at each point in time. For society today, as a complex, dynamic system, this is no longer a good approximation. The rate of change is such that the change within an individual's lifetime is significant, so that a decoupling of the dynamics of the individual from that of society is no longer appropriate.

At this point, let us just recall that what we are considering is the information of class 3—society's belief system. We are not dealing with the change to the rest of our knowledge base and the associated need for ongoing education throughout an individual's lifetime. That is very different; it is the response to a change that has taken place, whereas what we are considering is the individual's involvement in change itself, and education's new role is to prepare the individual for that involvement. It is not about acquiring information relevant to an expected future development of society, as it is impossible to know how society will develop; it is about acquiring and exercising an ability for critical thought, as discussed in Sect. 6.1.

This new role for education can also be seen as a shift in emphasis, from knowing to understanding. Until recently, the measure of intellectual prowess was the amount of knowledge a person could trot out at any time, and a profound thinker was often confused with someone who could produce appropriate quotations; now the knowledge and the quotations are readily available on demand. What we are experiencing is a change in our relationship to information, or knowledge. First came the accumulation of facts, knowledge being simply the knowledge of facts. Then came the application of facts, the knowledge of how to benefit from the accumulated facts, which developed through crafts to industry and the support of modern society. And now we are engaged in developing our understanding of the meaning of information, as something that emerges from the relationships between information items. This change is one of increasing complexity; our society has become so complex that only a holistic view of our knowledge and of the relationships contained within it will allow us to make correct judgements about its further evolution. And the means for operating on this system level of knowledge can no longer be the individual in isolation; it can only be the greatly increased processing power of individuals interacting through the free exchange of information—the *collective intelligence*. To give the individual the ability to participate effectively in this gigantic information processing venture by responding with a critical evaluation to the received flow of information is the new, additional task of education.

References

1. Durkheim E (1970) Education and sociology (trans: Fox SD). The Free Press, New York
2. Peters RS (1967) What is an educational process? (Chap. 1). In: Peters RS (ed) The concept of education. Rutledge & Kegan Paul, London

3. Passmore J (1967) On teaching to be critical (Chap. 12). In: Peters RS (ed) The concept of education. Rutledge & Kegan Paul, London
4. Lövlie L, Mortensen KP, Nordenbo SE (eds) (2003) Educating humanity: Bildung in postmodernity. Blackwell Publishing, Hoboken. J Phil Educ 36(3)
5. Biesta G (2003) How general can Bildung be? Reflections on the future of a modern educational ideal (chap. 4). In: Lövlie et al (2003)
6. Gur-Ze'er I (2003) Bildung and critical theory in the face of postmodern education (Chap. 5). In: (Lövlie 2003)
7. Reichenbach R (2003) On irritation and transformation: a-telelogical Bildung and its significance for the democratic form of living (Chap. 6). In: Lövlie (op.cit)
8. Herold B (2016) Issues A-Z: technology in education: an overview. Education week. Retrieved 13.01.2018 from http://www.edweek.org/ew/issues/technology-in-education/
9. Selwyn N (2016) Are we being too quick to embrace technology in education. Radio National. Retrieved 13.01.2018 from www.abc.net.au/radionational/programs/futuretense/are-we-being-too-quick-to-embrace-technology-in-education/7211366
10. Hanushek EA, Wössmann L (2010) Education and economic growth. In: Peterson P, Backer E, McGaw B (eds) International encyclopedia of education, vol 2, pp 245–252
11. Habermeier H-U (2007) Education and economy—an analysis of statistical data. J Mater Educ 29(1–2):55–70
12. Kim YJ, Terada-Hagiwara A (2010) A survey on the relationship between education and growth with implications for developing Asia. ADB Economics Working Paper Series No. 236
13. Solow R (1956) A contribution to the theory of economic growth. Q J Econ 70(1):65–94
14. Matthews J (2013) The educational imagination and the sociology of education in Australia. Aust Educ Res 40(2):155–171

Chapter 7
Political and Economic Factors

Abstract In addition to technology and education, politics and economics are significant factors determining the environment in which the model of society as an information-processing system operates. A novel picture of the economic influence on the individual is developed in terms of income level and inequality, and the influence of economics on the information exchange is discussed briefly. The most significant aspect of the political system, as far as information is concerned, is the party structure and the inevitable transformation of the party into a business, with the accompanying branding in the form of ideology.

As an introduction to this chapter, we note that, while the economic and political factors are treated in separate sections, there is a strong overlap or interaction between the two. But there is also an increasing strain in the relationship between them, and a change in the balance between them, due to the combined effects of globalisation and the concentration of wealth, giving rise to what has been called the Transnational Capitalist Class (TNC) [1]. The TNC is not interested in national concerns; its focus is on achieving a good return on capital, without any distractions created by party politics and inter-state squabbling. The national political elite, on the other hand, want to preserve their power—politics as a different form of business, performed by the party apparatus, whether in a representative democracy or in a one-party state. In the sense of "who pays the piper gets to call the tune", it is perhaps not completely inconceivable that one day the TNC might take over the management of the world, running it as a giant corporation; achieving by economic means what the UN failed to achieve by political means.

E. W. Aslaksen, *The Stability of Society*, Lecture Notes in Networks and Systems 113, https://doi.org/10.1007/978-3-030-40226-6_7

7.1   Economic Factors

7.1.1   Scope of the Considerations

At the beginning of this section it is perhaps worthwhile to briefly remind ourselves what this essay is about. It is about contributing to our understanding of the process— the operation of the collective intelligence—that drives the evolution of society, as reflected in the changes to society's belief system and, in particular, to the stability of this evolution It is not about contributing to our understanding of society per se, as in a description of what society consists of, or how it functions. And so, when we now consider economic aspects, they are not how an economist understands economics, as in fiscal and monetary policy, labour and capital as means of production, and market forces. In this essay, we are viewing society as a giant information-processing system, so the relevant aspects of economics are those that influence the functioning of this system. And from the previous chapters we can recognise that the most appropriate way of identifying these aspects is indirectly, by how they influence the collective intelligence of that system, with its three components:

- the individuals, with their information-processing capabilities;
- the interactions between the individuals, mediated by information technology; and
- the identities of the individuals.

The individual's identity is formed through interaction with the individual's environment; first by parents and siblings, then by friends, by formal education, and, finally, throughout the rest of the individual's lifetime by interaction with other individuals through various channels and by what we characterised as discovery. All of these activities are influenced by economic factors, but the major one is education, which was the subject of Chap. 6. The other two components are treated in the next two sections.

7.1.2   Sustaining the Individual

The first of these components—the living component—is the one most directly influenced by the economy, in the sense of being sustained by it, and the ability of the individual to perform its information-processing functions is strongly related to the most-often presented measure of the economy, the *per capita* Gross Domestic Product. Strongly, but not exclusively: If one person is able to have the best education, access to all possible sources of information, and spend all its time thinking, while everyone else is kept in total ignorance and just slaves away, the average information-processing capability will be very low, but if that one person is rich enough and owns all the production capacity, the average *per capita* GDP could be reasonably high. This example is, of course, a utopian extreme, but it illustrates that inequality must be an important economic parameter when it comes to the influence of the economy on the collective intelligence and the ability of the individual to participate in the social discourse, and we can expand this illustration by the following simple model.

Consider a society with n_0 members (to be defined), and form them into a set ordered by increasing income, $y(n)$, $n = 1, ..., n_0$, and $y(n + 1) \geq y(n)$. For large values of n_0 we can treat the variable $x = n/n_0$, $0 \leq x \leq 1$, as a continuous variable, and the income distribution function $y(x)$ as a continuous, monotonically increasing function. The cumulative income distribution, $Y(x)$, is given by

$$Y(x) = \int_0^x y(x')dx', \tag{7.1}$$

and we shall denote the cumulative total, $Y(1)$, by β. The Gini coefficient, G, is given by

$$G = 1 - \frac{2}{\beta} \int_0^1 Y(x)dx \tag{7.2}$$

Let us now, for the purpose of this illustration, choose the simplest form of $y(x)$,

$$y(x) = y_0 x^\alpha, \quad \alpha > 0,$$

where, as will become obvious, we have to exclude the case $\alpha = 0$, which corresponds to complete income equality. We then have

$$Y(1) = \beta = \frac{y_0}{1 + \alpha}, \quad or \quad y_0 = \beta(1 + \alpha); \tag{7.3}$$

and

$$G = \frac{\alpha}{2 + \alpha}. \tag{7.4}$$

To make the connection with our model of society, we assume that the ability of the member to participate in the collective intelligence is directly proportional to the income, up to an income y^*; for incomes above y^* the ability to participate does not increase. The income y^* may be identified with what is called the *living wage*. There is a corresponding value of x,

$$x^* = \left(\frac{y^*}{y_0}\right)^{\frac{1}{\alpha}}$$

which shows why $\alpha > 0$. The collective intelligence, CI, as a fraction of its maximal value y^*, is the cumulative value of this participation,

$$CI = y_0 \int_0^{x^*} x^\alpha dx + (1 - x^*)y^* = \beta(x^*)^{\alpha+1} + (1 - x^*)y^*. \qquad (7.5)$$

If we now make y^* the unit of monetary value, so that CI, y_0, and β are measured in units of y^*, then

$$CI = \beta(x^*)^{\alpha+1} + 1 - x^*; \qquad (7.6)$$

with

$$x^* = \left(\frac{1}{\beta(1+\alpha)} \right)^{\frac{1}{\alpha}}. \qquad (7.7)$$

The function CI(G, β) is shown in Fig. 7.1. In this figure, the dash curve indicates the limit of the model, arising from the condition $x^* \leq 1$, and some allowance has to be made for the coarseness of the calculation grid. But the basic message is unmistakeable (and not surprising): At low values of β, the better strategy for increasing the collective intelligence is to increase β, whereas for β values above the nominal income, y^*, it is much more important to decrease the inequality.

Now, to make this model more specific, and, as an example, relate it to an actual society—Australia, we need to define the nature of the "member" and that of β.

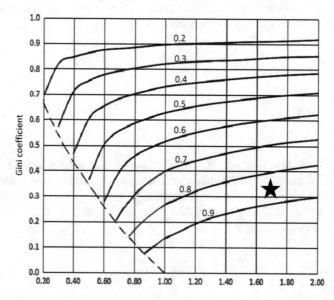

Fig. 7.1 Curves of constant value of the collective intelligence, as a fraction of its maximum value, in the plane spanned by the Gini coefficient and β, interpreted as the mean household disposable income relative to the living wage, y^*. The star indicates Australia's position in this evaluation

In the context of economic wellbeing, the Australian Bureau of Statistics (ABS) considers that the individual is best accounted for as a member of a *household*, and the disposable income required for a given level of wellbeing is dependent on the size of the household. But not proportional to the size; the *equivalent size* is given as 1 for the first adult (i.e. 15 years of age or over), then 0.5 for each additional adult, and 0.3 for each child. Then, for a lone person household, the *equivalised* income is equal to actual income; for households comprising more than one person, it is the estimated income that a lone person household would need to enjoy the same standard of living as the household in question, which is the income of the family divided by the equivalent size. Thus, in our little model, the "member" is the household, the income distribution is the distribution of the equivalised incomes of these households, and β is the ratio of the average disposable income to the living wage. From the ABS (6523.0—*Household Income and Wealth, Australia*, 2015–2016), we can extract the following data:

Mean household disposable income: $52,468
Median household disposable income: $44,356
Gini factor: 0.325

Views on how a living wage should be related to such income data varies greatly; one calculation, adopted by the Greater London Authority and reported in (https://www.economist.com/the-economist-explains/2015/05/20/how-a-living-wage-is-calculated), is that it should be 60% of the median disposable income, plus 15% for contingencies, which in the case of Australia would make it $30,606 and $\beta \approx 1.7$, which is shown as a star in Fig. 7.1.

The importance of the *distribution* of the aggregate parameters used in the neoliberal economic model has become an important issue in the last two decades or so, and has led to a significant body of work based on empirical studies, such as the seminal work of David Card and Alan B. Krueger, *Myth and Measurement: The New Economics of the Minimum Wage* [2]. An overview of this development, which constitutes a criticism of the neoliberal model, is provided by an article by Boushay [3].

7.1.3 The Interaction

Besides having the financial resources for sustaining the processing function, the collective intelligence depends on the interaction taking place between the individuals, which enables the averaging process—society's immune system—and provides the feed-stock for the generation of new information through reflection and critical thinking. While face-to-face interaction is still important, in particular in conveying sentiments, the flow of information impinging on the individual is increasingly mediated by technology, as we saw in the previous chapter, and we noted that the associated IT infrastructure represents a very significant investment. An increasing proportion of this investment is non-government, and accordingly it is looking to achieve a reasonable return. The search for that return is the major determinant of how economics influences the flow of information in society, and to develop an understanding of this, it is useful to consider three different types of communications networks:

a. The original person-to-person communications network, exemplified by the postal services, telegraph, and telephone (fixed, now also mobile). The basic business model was that users paid for the traffic generated, measured by weight and distance (postal service), words and distance (telegraph), and time and distance (telephone). In addition, if one wanted to have a private terminal, this was available at a rental or subscription cost, so that the total revenue generated by a user, u, was of the form $u = a + bq$, where q is a measure of the traffic generated.

For both the postal service and telegraph (which is practically extinct), $a = 0$. For the fixed telephone network one has a choice, in principle, between being a subscriber, in which case $a > 0$, or using a public phone, if one can find one, in which case $a = 0$. Today, in many countries, the main telephone network is the mobile network, which requires the user to have a terminal, so $a > 0$. A user must also purchase a plan (recurring or pre-paid), which effectively adds to both a and b, but with a pre-paid plan, this cost can be kept very low, so that the main economic barrier to basic telephone service (voice call) is the cost of the mobile telephone.

b. Broadcasting, or one-to-many networks, exemplified by newspapers, radio, and television, and the users of these networks are readers, listeners, and viewers. Here the business model is quite different, because the main part of the business is not the communications network, but the provision of the *content* to be *distributed* on the network. User-generated traffic is not a relevant concept—the relevant measure is the number of users multiplied by the time each user spends on consuming the content; this is the amount of the user's *attention* the network is able to capture. And the revenue does no come from the users, but from advertising, and depends on that amount of user attention. Hence, the business model essentially becomes a matter of optimising the difference between the advertising revenue and the cost of producing attention-grabbing content.

In the case of newspapers, there is no fixed cost; $a = 0$, and b represents the cost of single newspaper. The cost of accessing wireless (or free-to-air) services is the cost of the terminal device; i.e., a is represented by the cost of a radio or a TV set, and $b = 0$. However, except for some government-funded services, the user has to endure an amount of advertising.

c. Computer networks, networks that can carry any information as long as it is digitised; speech, text, music, graphics, videos, smell, touch—all that is required are the corresponding transducers connected to the computers at the end of a communications path. For the ordinary user, the only network of importance is the Internet, and from the user's point of view there are two groups of entities involved: the Internet Service Providers (ISPs) and the content providers (where content includes service). The former provide access to the network and certain communications services, such as email; the latter provide almost any content or service imaginable, including data bases (e.g. Wikipedia), voice and video communication (e.g. Skype, on-demand video), social media (e.g. Facebook and Twitter), games, booking services, dating services, blogs, chat rooms, and so on, with search engines (e.g. Google) providing the means for navigating through this mass of sites. The choice of ISP does not limit the size of the network; subscribers are subscribers to the Internet, only via a particular ISP.

The economic aspect of the Internet, as experienced by a subscriber, is the result of a combination of two business models; one for the ISP and one for the content provider. The business model for the ISP is very similar to that of the type a. network, except that with the increasing roll-out of free Wi-Fi, it is not necessary for an individual user to subscribe to an ISP at all (but, of course, the entity providing the free Wi-Fi needs to be a subscriber). A great deal of useful information and services are available for "free", but often with the acceptance of being subjected to advertising and, in many cases, having to register, which allows the advertiser to send emails to the user or pop-ups to the user's terminal device. However, many of these services and content sites also offer the same, or premium, content without advertising, but at a price, so that, if we consider that advertising takes up some of the available attention time, and that time is money, then advertising can be seen as an impost on the poor.

From this brief description of the main communications networks that mediate the interaction between the members of a society, we see that, aside from the fact that advances in information technology allow increasingly convenient and intense communication, our market-based economic system leads to inequality in what should now be a basic social good, along with education and healthcare. But in the context of this essay, with its emphasis on the collective intelligence and the role of interpersonal interaction, there is another aspect of the digital age that is of some importance, and it is the extent to which applications of information technology are reducing the time spent on interpersonal interaction. This issue first came to prominence with the appearance of TV in the majority of homes, and raised concerns about people, but in particular children, becoming passive viewers and neglecting such activities as hobbies, sports, and the social intercourse with family and friends. While the dire predictions raised in this regard were not realised, there are now two new contenders for the attention of the younger generation, and the first one is the various applications that go under the heading of social media. Most of these applications can, in principle be used to carry out an ongoing, two-way interaction, but in practice they are used overwhelmingly as a one-way channel of *posting* information, often in the form of opinions, but also just as a message, as in a photo. Feedback in the form of "likes" and "dislikes" does not count as constituting a conversation.

The second one is *gaming*. The published data on usage varies from country to country and between different sources, but the gaming industry is reported to be worth something like US$140 billion in 2018, and rapidly increasing, with such applications as esports and virtual reality games. Games are, of course, by their nature interactive, but what the player is interacting with is not another person, but a program that is responding according to fixed rules. It may be good mental stimulus to try and figure out what the rules are and then outsmart them, but it does not contribute to developing the identity or to the effectiveness of the collective intelligence. And there is, as with so many of the information technology applications, the opportunity to utilise games to present a particular ideology, sometimes openly, but more often covertly or even inadvertently.

The above presents one factor in the influence of economics on the interaction between the members of society; it is what we might consider a direct factor, in that it shows how the cost of IT services constrains a person's access to those services, and thereby to the information required for the operation of the collective intelligence. There is another, more indirect, economic factor that interferes with the information flow, and it is caused by the concentration of wealth in a very small proportion of the population. On a global scale, 47% of household wealth is owned by 1% of the adult population, and 150,000 people each own more than US$50 million (https://www.credit-suisse.com/about-us/en/reports-research/global-wealth-report.html), With this wealth comes control of a large part of the IT services, and with control comes the ability to modify the information in various ways, ranging from simply deleting unwanted information to actually producing "fake news". An added factor here is the degree of privatisation of the IT services. In many Western nations privatisation is part of the economic ideology, and the flow of information is slanted towards the interests of the owners; whereas in nations with stronger government ownership and control, such as in China, the information is modified to reflect government policy. In either case, the control of the IT services, together with the ease of modification provided by IT, pose perhaps the most serious threat to the operation of the collective intelligence. This realisation is nothing new; it was the subject of numerous publications in the 1980s, with perhaps the best known being the book by Edward S. Herman and Noam Chomsky, *Manufacturing Consent: The Political Economy of the Mass Media* [4] and the two books by Herbert I. Schiller, *Who Knows: Information in the Age of Fortune 500* [5] and *Culture, Inc.: The Corporate Takeover of Public Expression* [6].

7.2 Political Factors

7.2.1 The Party System

In one form or another, the concept of the political party is central to the operation of most nation-states. Whether a one-party state or a representative democracy, the political power is wielded by a relatively small group of people who manage the affairs of government in much the same way as the directors and senior executives manage a corporation. In both cases the members of the society constitute the market;

in both cases the business strategy is a combination of persuasion and an adaption to market demands; and in both cases the objective is survival against internal and external competitors. And in both cases the "will of the people", which in our model is represented by the adaptive action resulting from the operation of the collective intelligence, is expressed in a passive fashion, by either accepting or rejecting what is offered.

It is here that there is supposed to be a significant difference between a one-party state and a representative democracy; the latter offers a choice between several alternatives, as well as the ability to propose new alternatives, whereas in a one-party state there are no alternatives, and the choice is, at best, between acceptance and passive rejection. In practice, there is actually not that great a difference between the two; in both cases, the governing party, or coalition of parties, prosecute the daily business of managing the state in accordance with a pragmatic evaluation of the existing circumstances based on their own interests, with public opinion just one of many factors—and the one most easily manipulated. And in both cases there is the same undercurrent in the population of frustration with the lack of a more direct and effective participation in the decision-making process; people feel they are not being heard. The level of frustration will, of course, depend on the extent to which the regime meets the basic needs of society; i.e., on the standard of living, but once these needs are met, and the level of education and leisure time increases further, the level of frustration rises again. In single-party and authoritarian states it is simply suppressed; in the liberal democracies it is exploited in support of minority grievances related to such issues as nationality, race, gender, and religion. To the extent that such actions are successful—possibly even resulting in a change of government—this display of people-power may appear as a win for democracy, but in reality it is circumventing the collective intelligence and represents a fluctuation away from the optimal path.

The multi-party and, in particular, the two-party system is also inefficient, at least in practise, if not in its ideal form, as the competitiveness and antagonism of the election campaign is carried over into the operation of the legislature between elections. As an example, in Australia much of the time during the sitting of Parliament is spent on childish finger-pointing (literally) and point-scoring, and the time between sittings is spent on building the base in the constituency for the next election. The productivity, in terms of developing well-considered responses to the many issues and problems arising, is much lower than it could and should be; the only country that approaches the ideal is Switzerland.

A single-party system should, again in principle, be able to avoid the inefficiencies of the multi-party system. And if we define a democratic system as one that realises the will of the people, then a one-party system could, in theory, be just as democratic as a multi-party system, with the party organisation providing the structure for the upward movement of that will from the grass-roots to the government. However, in practice the single-party system is more susceptible to the establishment of a ruling class that squanders much of what should be a gain in efficiency in realising the will of the people on maintaining its own privileged position.

In *The Social Bond* a proposal for abolishing the party system and strengthening the individual's direct involvement in the political process was put forward, based on current advances in IT. There is no real technical or cost reason why such a scheme could not be implemented; the reasons are, firstly, that the party system has developed into quite a substantial business, and secondly, that the competition between parties deflects attention and effective action away from privileged positions in society. The reality is that there are significant and powerful sections of society that have no interest in promoting a more inclusive political system, and that see "the will of the people" as an obstacle in pursuing their interests. And in most cases this is not a matter of reasoning based on a sound understanding of the nature of society as a complex system, but simply an ideology, which is the subject of the next subsection.

7.2.2 The Role of Ideology

When we, in the previous section, said that the daily actions of all governments are the result of pragmatic assessments of the situations confronting them, then 'pragmatic' refers to accepting the situations as they present themselves and adjusting the actions accordingly in order to achieve the best possible outcome in each case. Consequently, appropriateness or otherwise of the actions can only be judged relative to the definition of 'best'. In this essay, 'best' is defined, not in terms of specific characteristics, such as freedom, wealth, equality, or rule of law, but in terms of a process: 'best' is whatever results from the operation of the collective intelligence.

In Western societies it is generally assumed that our definition of 'best' has until now most often been synonymous with the concept of a *liberal democracy*; a concept that arose out of the Enlightenment ideas of liberty and equality, and which, according to Wikipedia, is characterised by the following principles:

a. elections between multiple distinct political parties;
b. a separation of powers into different branches of government;
c. the rule of law in everyday life as part of an open society;
d. a market economy with private property; and
e. the equal protection of human rights, civil rights, civil liberties and political freedoms for all people.

A society realising these principles, taken at face value, appears to be a reasonable (but not the only) candidate for the "best" society. However, it is the application of these principles—their realisation in the everyday operation of the society—that is relevant to any assessment of "best". The electoral process is ineffective if the parties differ only in detail; both the separation of powers and the rule of law only become issues in a dynamic society; market economy and private property mean nothing until their extent and limitations are defined; and the items in e. are open to a wide range of interpretations. The overriding characterisation of Western societies is in terms of an *ideology* that determines the operation of all the principles in the above list; it is an ideology that has emerged as a result of two historical developments: Christianity and the expansion of European power and the creation of what we now call the West. The latter was discussed briefly at the end of Sect. 3.4, and resulted in a perception of the value and importance of people as individuals rather than as members of society, and devaluing such concepts as social responsibility and solidarity.

By promising everlasting life after death, and making belief the determining factor for achieving it, Christianity made inequality and exploitation in this life relatively irrelevant. Indeed, without inequality, how could there be any Christian charity? Suffering and death in this life, as was the fate of millions of natives, was a small price to pay for eternal life; better to be a slave as a Christian than free as a heathen. Through this relationship between faith and the qualities of daily life, the administration of the faith—the Christian church—arose as a powerful organisation in its own right and as both a collaborator with and competitor to the State, as illustrated by the power struggle between Pope and Emperor during the Investiture Controversy around 1100 AD. In the end, control over social life was shared, with the Church controlling two fundamental aspects of every person's life—sex and death—while the State controlled the economic aspects. The Church was given favourable treatment in economic matters, and the State was given more or less automatic absolution for its sins. And in order for the Church to be an effective partner, the original emphasis on faith was augmented and partially replaced by an emphasis on the individual's duty to contribute to the economy through diligent work and the accumulation of wealth; exemplified by what is termed "the protestant work ethic" (not that Catholics are any more adverse to accumulating wealth). It is this particular amalgamation of religious dogma with the extraordinary economic circumstances in the Western world in the last five hundred years that has created the neoconservative ideology.

When referring to societies embracing this ideology it is usually not as neoconservative or capitalist (the latter has a bad taste, like communism), but as liberal democracies. This terminology cloaks what is essentially an economic ideology with adoration of private wealth and conspicuous consumption as something nobler; as the realisation of an intrinsic human aspiration, like motherhood. Conversely, labelling a society or its behaviour as undemocratic immediately characterises it as deplorable, without the need for any further elaboration or analysis. The meaning of democracy is literally "rule by the people", and Wikipedia states "Democracy is a system of government where the citizens exercise power by voting.". There is an implicit assumption that the power exercised by voting is the only power, and that power is therefore distributed equally over all citizens. In reality, in all democracies there are numerous other avenues to exercising power, and if we could measure the power exercised by each person and create a Lorenz curve, as one does in order to determine the Gini coefficient, I would expect the curve for power to be very similar to the curve for wealth, and the power Gini coefficient to be correspondingly large.

But besides the many questions about practical aspects of this definition, we must first ask: What is the purpose of government? Only then can we judge the efficacy of democracy as a system of government. And to answer that question, we must recognise the special nature of society as a system of interacting individuals. In all other systems, the system has a purpose defined by its relationship to something outside itself. In the case of a telephone system, the purpose is to provide a communications service to its users; in the case of a car, the purpose is to provide certain services (transport, prestige, etc.) to the owners; in the case of the education system, the purpose is to provide knowledge and understanding to the students (in the widest sense); and so on. In the case of society, the purpose can only be defined in terms of the service it provides to its members; i.e., to the very elements that make up the system. It is as if the purpose of a car would be to make the engine run better.

Therefore, to answer the question, we fist have to define the purpose of society. One approach is to see the purpose in terms of supporting the individual in satisfying its needs, and with reference to Maslow's hierarchy of needs [7], we might define a measure of how well a society meets its purpose as the degree of self-actualisation reached by the individual, averaged over all the individuals in society. The extent to which the lower needs have to be satisfied in order for effort to be allocated to self-actualisation; i.e., the relative strengths of the needs, will vary from person to person, from society to society, and with time within each society, and hence it is unlikely that a single form of government will be appropriate for achieving the most effective distribution of society's resources over these five levels for all societies. And there does not seem to be any rational argument for why Western liberal democracy is an optimal form of government for any society; it is an ideology that has emerged as a result of the special circumstances experienced by the West in recent times.

The dominance of the neoconservative ideology in the West reached its peak in the 1980s, personified by Ronald Reagan and Margaret Thatcher, and claimed its ultimate vindication in the collapse of the Soviet Union in 1991. In the words of Francis Fukuyama: "That is, the end point of mankind's ideological evolution and the universalization of Western liberal democracy as the final form of human government." [8]. However, this exuberance glossed over many of the problems arising from different cultures and religions. In the book *Spectres of Marx*, Jacques Derrida criticised the Fukuyama view, provided a list of the problems in the form of "10 plagues of the capitalist system", and pointed out some of their effects:

> For it must be cried out, at a time when some have the audacity to neo-evangelize in the name of the ideal of a liberal democracy that has finally realized itself as the ideal of human history: never have violence, inequality, exclusion, famine, and thus economic oppression affected as many human beings in the history of the earth and of humanity. Instead of singing the advent of the ideal of liberal democracy and of the capitalist market in the euphoria of the end of history, instead of celebrating the 'end of ideologies' and the end of the great emancipatory discourses, let us never neglect this obvious macroscopic fact, made up of innumerable singular sites of suffering: no degree of progress allows one to ignore that never before, in absolute figures, have so many men, women and children been subjugated, starved or exterminated on the earth. [9]

In the context of this essay, the central feature of this ideology is its emphasis on a person as an individual relative to a person as an element of society. It is the individual's faith that is the determinant in achieving salvation and eternal life; the only relationship is between the individual and God; and a person's achievement is expressed predominantly in the accumulation an individual ownership of wealth. Society, as the entity created by individuals through their interactions, does not feature in this narrative. The opposing ideology, socialism, emphasizes the importance of society and the mutual dependence of society and individual—the Social Contract—and recognises the importance of the state as the realisation of the structure arising out of the interaction between individuals. And as the interaction increases, it is acknowledged that society becomes increasingly dynamic and increases the demands on its control. One way to look at this is that, as the level of interaction increases, the number of possible configurations increases and, as an expression of the Law of Requisite Variety [10], which we mentioned in Sect. 3.6, for society to remain stable, the control system, i.e., government, needs to develop accordingly and have an increasing number of options at its disposal. The neoconservative war cry of "small government" is a reflection of an ideology removed from reality; government needs to be complex, dynamic, effective, and efficient; size is not, in itself, a significant measure.

References

1. Robinson WI (2014) Global capitalism and the crisis of humanity. Cambridge University Press, Cambridge
2. Card C, Krueger AB (1995) Myth and Measurement: the new economics of the minimum wage. Princeton University Press, New Jersey
3. Boushay H (2019) A new economic paradigm. Democr: J Ideas 53
4. Herman ES, Chomsky N (1988) Manufacturing consent: the political economy of the mass media. Pantheon Books, New York
5. Schiller HI (1981) Who knows: information in the age of fortune 500. Ablex, Norwood
6. Schiller HI (1989) Culture, Inc.: the corporate takeover of public expression. Oxford University Press, Oxford
7. Maslow AH (1943) A theory of human motivation. Psychol Rev 50:370–396
8. Fukuyama F (1992) The end of history and the last man. The Free Press, New York
9. Derrida J (1994) Spectres of Marx: the state of the debt, the work of mourning and the new international (trans: Kanuf P). Routledge, Abingdon
10. Ashby WR (1956) An introduction to cybernetics. Chapman & Hall Ltd., London

Chapter 8
Stability

Abstract This final chapter focuses on the motivation for this essay; the concern regarding the stability of the evolution of society. After some general comments on the nature of social stability and its expression in the model, the current situation is considered from two points of view—the stability of nation-states and the stability of the world as a single society. In the former case, while instabilities (fluctuations) may have devastating effects on the individual nation-states, they are highly unlikely to have such an effect on society as a genus. The latter case, however, given the absence of any external threat or competition, is of greater concern, and while a catastrophic fluctuation may be unlikely, its probability is by no means negligible.

8.1 The Nature of Social Stability

In this final chapter we take a closer look at the issue that motivated this essay—the stability of society—based on our model of society as an information-processing system and on our understanding of the factors that influence its operation and evolution, as it was developed it the preceding chapters.

According to the Canterbury English Dictionary, stability is "a situation in which something is not likely to move or change". Depending on the nature of the "something", the concept takes on a more specific meaning. For example, the stability of a structure, such as a building or a bridge, refers to its ability to withstand external forces, as might be experienced in an earthquake or a cyclone. In the case of a control system, it is its ability to maintain the controlled parameter close to its set-point value under varying conditions. And in materials science it is the ability of a substance to remain unchanged under changes to its environment. When we now want to consider the meaning of stability in relation to society, it can obviously not mean that society is unchanging, as by its very nature society is continuously evolving. If we view this evolution as the result of a process, then perhaps stability could mean that this process remains unchanged? In the case of an industrial production process, stability means that the characteristics of the product remain within certain limits; but this type of stability cannot apply to the process driving the evolution of society, as the changes that took place a thousand years ago were quite different to those

E. W. Aslaksen, *The Stability of Society*, Lecture Notes in Networks and Systems 113,
https://doi.org/10.1007/978-3-030-40226-6_8

taking place today, both with respect to the rate of change and to the substance of the changes. So, if social stability can mean neither no change nor a prescribed change, including progressing toward an end state, is the evolution of society just the result of a random process, in which case the concept of stability has no meaning?

Both our intuition and the historical record would indicate that this is not the case. From the early examples of society, say, in the form of nomadic tribes, until our current nation-states, there has been a desire to regulate change through a set of rules and customs, and what was considered normal changes at one point in time, e.g., intertribal warfare, became undesirable fluctuations at a later date, when the tribes had been integrated into a larger structure, such as a kingdom. At each stage, there has been an understanding of what is an acceptable change and what is an unacceptable change—a fluctuation away from the desired evolution of society. The frequency and severity of such fluctuations constitute the manifestation of the instability of society.

In Sect. 3.5 we introduced the concept of dividing the activities taking place in society into two cycles, with all the activities we associate with daily life, and described by economics, politics, ethics, religion, and, more generally, by social science, making up the economic cycle. In our model of society as an information-processing system, there is a part of the information processing associated with the economic cycle; let us call it the *secondary* process. This process includes the research, innovation, and increase in experience that drives the evolution of the economic cycle—the increase in GDP, the new applications of technology, and even new works of art, but it also includes such changes as wars, economic crises, and the like; changes that we would consider *fluctuations* of the evolution of the economic cycle. The severity of a fluctuation is measured by such characteristics as the deterioration in the satisfaction of basic needs, loss of life, and destruction of the built and natural environments. The *stability of society* is a measure of the degree to which such fluctuations are limited in frequency and severity; it does not mean that the economic cycle is not evolving.

Underlying the actions of the economic cycle is a belief system or an ideology; we might say that the actions and evolution of the economic cycle *represent* this belief system, and the stability of society is a characteristic of this belief system. The view put forward in this essay is that this belief system is itself evolving, and that there is another process—the *primary* process—that drives the changes to this belief system. The primary process evaluates proposed changes to the secondary process, and the ideal version of this primary process is what we, in Chap. 4, called the *collective intelligence*. It is the belief (and it can only be a belief) supporting the work in this essay that the operation of the collective intelligence will provide the most stable evolution of society; i.e., of society's belief system.

The core of the collective intelligence is the individual's ability to survive, to determine to what extent a new situation is relevant to survival and what adaptive action is most likely to increase the probability of survival. This ability is inherent in every living creature; its power and sophistication is appropriate to the creature's environment and to its ability to take adaptive action. It is not something that at present, if ever, can be further defined or detailed; its existence is a matter of personal interpretation of evolution. It may perhaps best be viewed as the defining feature of *life*; life is about survival. In humans, this ability is highly developed and based on a learning process, so that as society evolved in the level of integration and complexity, the understanding of what constitutes survival and the criteria for achieving it evolved accordingly; mostly through a process of trial and error, coupled with reflection, and transmitted from one generation to the next through the process described in Sect. 6.2.2.

However, the power of the collective intelligence as the guiding process for the evolution of society comes from the interaction between individuals—the public discourse that leads to a consensus and the potential for action—and this interaction can be defined and parametrised in an indirect manner by identifying the characteristics of society that determine the operation of the collective intelligence. This is what was proposed in *The Social Bond* and outlined in Sect. 4.3, where we defined a single composite parameter that represented the restrictions society places of the operation of the collective intelligence; in the absence of any restrictions, society will evolve along a trajectory that minimises fluctuations. In the next two sections we shall examine some particular restrictions and their effects; here we first look more closely at restrictions and fluctuations in general.

Due to the activities of individuals, there will always be a large number of ideas and proposals for changes to society's belief system. If there are no restrictions, so that the collective intelligence is operating freely, the overwhelming majority of these will not have any significant impact, mainly because they are seen as beneficial only by a small part of the population, but some of these initiatives will have a more significant impact within a local part of society and a limited timeframe; they are the fluctuations associated with an *equilibrium* state of the belief system. Only a very small fraction survive the assessment and consensus-building process of the collective intelligence, and these are the changes that constitute the evolution of society—a slow, continuous change of the equilibrium state. (Note the similarity with the dynamics of equilibrium thermodynamics.) Fluctuations will be of limited duration due to the averaging effect of the collective intelligence; they are to society as infections are to the human body, and the collective intelligence plays the role of a social immune system.

This picture changes if there are restrictions on the operation of the collective intelligence, which is, of course, the case in all societies today. The primary process is then no longer identical with the collective intelligence, and while there are various reasons for this, depending on the particular society, they are basically of two types: One, arising from the lag between the problems arising from the technology-driven changes to society and the development of their solutions; this is the situation depicted in Fig. 4.6. Two, arising from the resistance to any change on the part of society benefiting most from its current state. It is this latter type of actions that are most relevant to our concerns about stability, because, while the economic cycle may be in an equilibrium state relative to the operation of the secondary process under the current belief system or ideology, the overall evolution of society, under the combined primary and secondary processes, is diverging further and further from the ideal path, with an increasing probability of a catastrophic fluctuation. This divergence is reflected in a corresponding divergence from the uniform society assumed in our model—a segmentation of the society that appears as one segment imposing a limitation on the free generation and flow of information. Now, when we speak of the collective intelligence, of its operation, of the restrictions placed on it, and of social stability, this is always in relation to a particular society. Societies range in size from a family to the world society, and the conditions under which the collective intelligence operates vary accordingly. And when we treat a society as a stand-alone entity, that is always a simplification, in that any society, with the exception of the world society, is embedded in a larger society, as an element of that society. In the last two sections of this essay I look at two societies—the nation-state and the world society—that are of particular interest in the context of stability, not least because the relationship between them—the "embedding" of nation-states in the world society— is unique due to the fact that the world society is "the end of the line", so to speak, in the sequence of societies, as was shown in Table 1.2.

Before leaving this section on general aspects of social stability, we should note the rising visibility in the last 25 years of the central role of religion in politics and in the underlying ideology. This is, of course, nothing new; the priesthood and its organisation has been a major component of any state since the beginning of historical times. But now, perhaps in reaction to the long domination of Christianity, all the three other main religions—Islam, Hinduism, and Buddhism—are showing signs of reasserting their role in politics, as evidenced by the persecution of the Rohingas and rising militant Buddhism in Myanmar; the increasing emphasis on Hinduism in Indian politics; the blasphemy conviction of the mayor of Jakarta and a shift towards a more strict interpretation of Islam, with a leading cleric as the running mate of Widodo in the recent elections in Indonesia; the strength of the Taliban in Afghanistan; the call for the introduction of Sharia law in some African states; and the rise of such a movement as Islamic State. And even in the West, the influence of religion seems to be on the rise, as evidenced, for example, by the current discussions about religious freedom in Australia and about abortion in the state of New South Wales, the election of an ultra-conservative Catholic as president in Brazil, and the religious component of nationalism in Eastern Europe.

8.2 Intranational Stability

Nation-states vary greatly in size, from China, with a population of almost one-and-a-half billion, to island states in the Pacific with only a couple of thousand inhabitants, and, correspondingly, they vary greatly in structure and complexity. So, when we now look at *the* nation-state, as a typical representative of the society of interest, it will be a better representative of some than of others; it will generally correspond to a developed nation, and in the first instance, to what goes under the label of a representative democracy.

The single most significant characteristic of the current evolution of society is— despite a certain government redistribution effort and (theoretically) progressive tax- ation—the increasing importance of capital and the associated inequality in both wealth and income, and thereby in power. Now, the situation where a small section of the population dominates the activities of society and reaps the benefits of them is nothing new; that has been the case since the beginning of historical times, with ruling families and nobility. But there has also been an unmistakeable trend towards greater equality and participation in the affairs of society; first in the form of a merchant class, then the trades through their guilds, then the manufacturers and the supporting professions, and finally the workers themselves. It can be seen as an increasing importance of the collective intelligence; the significance of the current development is that it is defying this trend.

The driver of this development is the predominant ideology in the Western democracies; it goes under various names, such as neoliberalism, neo-conservativism, or simply capitalism, but its main tenets are private ownership, a self-regulating, profit-driven market economy, and small government with user-pays for all services. Besides being an inherently unstable system, due to exponential growth, there are two immediate effects of this ideology: One is the pressure on finding new investment opportunities for the accumulating capital, reducing the individual to a form of consumption slavery. The other is suppressing any opposition to what is a highly non-equilibrium situation by subverting the political process; what appears as representative democracy is increasingly simply a reflection of economic reality. We are thus faced with an amazing development—a system that promotes a shrinking proportion of the population, and that is being criticised and mistrusted by an increasing part of the population, is still preferred by the majority.

There are a couple of reasons for this development, and one is the fact that, despite many grumblings, life is pretty good for the majority of the population. Taking Australia as an example, most of us have a place to live, enough to eat, adequate clothing, and the basic modern conveniences, such as a washing machine, refrigerator, TV, and mobile phone, and while many might wish they could afford more (due in large part to advertising), it is not enough to send us to the barricades. Another reason is that not only does our consumer society provide numerous distractions, but the barriers to contributing effectively to change are surprisingly high. One might think that with the means of communication at our disposal, it would be relatively easy to be heard, and indeed, there is nothing stopping us from posting our opinions and insights on social media and on websites, where they are freely available to everyone, but it is exactly this ease of posting that is reducing the Internet's effectiveness as a forum for discussion and development of ideas as part of the political process. And anyone that has tried to contribute through membership in the local branch of a political party will know that this is a very time-consuming and inefficient approach, and not a realistic option for most people with a busy professional life. The result is that an effective involvement in politics is really only possible for those who want to make it their career, which brings us back to the political party as a business.

We might think that running political parties as businesses should not be such a bad thing; after all, many companies are run very successfully by providing products to society's members in a competitive market. The difference is that in the political arena the competition is missing—there is no other process competing with the party system. The "competition" between Liberal and Labor, or between Republican and Democrat, has little effect on the party as a business, and so there is little incentive for change. What can change, and is slowly changing, are the policies of the parties; these changes are partly in response to the demands of the electorate, but also in response to special interest groups and, significantly, in response to the wishes of the TNC—all evaluated relative to the party as a business. Given this, it is unlikely that Western societies will experience catastrophic internal fluctuations, such as happened in the French and Russian revolutions, in the foreseeable future.

If we look beyond the liberal democracies of Western nations, we see a very different picture. With the exception of China, which has the longest continuous existence of any nation-state, most of the other nations were created either by the colonial powers, as in the case of Middle-Eastern and African nations, or by the beak-up of forced configurations, such as in the case of the Ottoman empire, the Soviet Union, and Yugoslavia. These states often have significant internal ethnic and religious strain, and little experience with democratic government, with the result that there has been a rising number of intrastate armed conflicts since the end of WW2. A recent book, *On Building Peace: Rescuing the Nation-State and Saving the United Nations*, by Michael von der Schulenburg [1] describes and analyses these intrastate armed conflicts in great detail, based on both on-site experience and a long career in peace-building efforts by the UN. His analysis shows how the current legal and institutional framework for maintaining peace is inadequate, as it is focused on avoiding and containing interstate conflicts, and he proposes how the UN could be restructured and revitalised to be effective on the intrastate level. But equally important is the strengthening of the nation-state as the building block of the global society, something that has come under attack as a result of a misconstrued understanding of globalisation.

These intrastate conflicts have been—and still are—exacerbated by two factors: the interests of the arms industry in promoting them, and the competing interests and power struggles of the major powers, with the result that many of these intrastate conflicts have taken on the form of proxy wars between the major powers. In particular, they are symptomatic of the West endeavouring to hold on to its dominant position, both economically and ideologically, and therefore, devastating as some of these conflicts are, such as the current one in Syria, they are in themselves not a threat to the stability of evolution. That threat arises on the international level, which is the subject of the next section.

8.3 International Stability

If we consider the view of evolution presented in the first part of this essay, with the evolution of the *genus* society as the current stage in this general process, we would expect the next *species* to emerge to be a world society, where the operation of the collective intelligence is extended to encompass all the individuals in the world. As in all other societies that evolved, this extended reach of the interaction between individuals would result in a further structuring, and the world society would be defined by a new level of governance, perhaps mirroring the national executive, legislative, and judiciary triple function structure. Within this structure, each nation-state would be able to develop for the benefit of its population and to be a valuable contributor to the wold society, just as individuals should be able to do within a nation.

The first attempt at such a structure was the League of Nations, founded in 1920, with the main objective of preventing war between nations. It was never very effective, not least because the US never joined, and was replaced by the United Nations (UN) after the end of the Second World War. With its numerous agencies, programs, and affiliated organisations, the UN has been, and is, quite successful in promoting and sustaining international cooperation in various fields, and as such must be see an initial, but substantive part of such a world-wide structure. However, with regard to maintaining stability between its member states, the UN was completely sidelined by two situations, as has been very well discussed and documented in the book by Michael von der Schulenburg referenced earlier—*On Building Peace* [1]. The first was the Cold War arising as a result of WW2, with two superpowers—the Soviet Union and the US—confronting each other in a precarious balance of nuclear weapons power known as Mutual Assured Destruction (MAD). Correspondingly, the world became divided into two spheres of influence, with numerous skirmishes, both covert and overt, on the fault line between them, with the two most serious ones being the Korean and Vietnam wars, and within this scenario there was clearly no scope for effective UN action. The second arose after the fall and disintegration of the Soviet Union, with the US emerging as the sole, and greatly strengthened, superpower. Emboldened by this success, the US set about providing "peace and order" on its own terms, while aggressively expanding its sphere of influence through such means as NATO, an organisation that should have been disbanded following the end of the Cold War. Again, this state of affairs made it practically impossible for the UN to have any decisive influence on the relationship between nations.

In Sect. 3.4 it was suggested that the wealth and dominant position of the West was, to a significant extent, due to the subjugation and exploitation of a largely defenceless third of the world over the last 500 years or so. This phase of the evolution of the world society came to an end around the middle of the last century, in part as a result of two world wars that devastated much of the West, but from which the United States eventually arose as the sole superpower. The nations that had, prior to 1900, made up what would be considered the West became politically and militarily subservient to the US, and much of the rest of the world was brought under US domination by economic means (sanctions or grants) or by direct military intervention (as in Iraq), as well as by an extensive repertoire of proxy wars and covert operations; a number of the latter, such as those in Guatemala, El Salvador, and Nicaragua, are documented in the book *How the World Works*, by Chomsky [2]. Thus, the world society is currently in a highly non-equilibrium state, where the supremacy of the US-led West is maintained by military and economic force against what in this essay is presented as the natural evolution of the world society, and such a non-equilibrium situation always carries with it the possibility of a collapse, if the forces supporting this balancing act should no longer be able to maintain it.

However, it is not the existence of the current non-equilibrium situation that is of greatest concern; if there was a will to let it settle gradually back into an equilibrium state that would be possible; albeit not without some significant adjustments for many nations. The greatest concern is the US effort to maintain and strengthen its hegemonial position by polarising the world into two camps: "Either you're with us, or you're against us." This was already a clear strategy following WW2, with the Soviet Union and communism defining the polarisation; any nation that only wanted to be itself, such as Vietnam, soon got to feel the consequences. With the collapse of the Soviet Union and its influence over Eastern Europe and Central Asia, one could have expected the polarising momentum to have faltered. But Russia could still be made into a threat justifying the strengthening of the Western alliance and the use of NATO to push eastwards and enforce US policy. Following the end of the Cold War and the dissolution of the Warsaw Pact, one could have hoped for a return to a world of mutually interacting nation-states, based on non-exclusive bilateral relationships under the umbrella of a world-wide framework, anchored in the UN. But this was not a view compatible with the US imperial vision of itself.

And now a new polarising factor has arisen—China—which has become the subject of the same kind of aggressive containment policies previously reserved for Russia. But while Russia is still a major power militarily and through the size of the area within its borders, it is no comparison with China when it comes to economic power, with a GDP of only US$1.6 trillion as compared with China's US$13.6 trillion [3]. From both its size—population and economy—and its history, it is evident that China is currently the main challenger to the hegemonial status of the West, and consequently I will limit my further discussion of the risk associated with the current imbalance to the relationship between the West and China. The role of other nations will feature only to the extent that they are drawn into the contest between these two parties.

The relevant information includes, of course, information about the current situation—the state of the two parties, their relationship, and their behaviour in the world society, but it also includes information about the recent history of China and its relationship to the West, as this history has a significant influence on what we would call the identities of the two parties. The following is mainly my severely abridged and edited version of the information provided in the second edition of *A History of China*, by Eberhard [4].

Leading up to the end of the eighteenth century, China had been developing into an increasingly prosperous and well-managed nation-state under nominal Manchu rule, but in reality administered by a Chines bureaucracy. External relations were very limited; external trade was only a few percent of the internal commerce, and the activities of foreign traders were severely restricted. The social structure was composed of the ruling Manchu, with their emperor and military garrisons, an administrative bureaucracy, a class of landholding gentry, a small and weak middle class of traders and manufacturers, and then the greater part of the population (>80%) as tenant farmers. But the time around 1800 became a turning point in the development of the Chinese society and the start of an accelerating decline. There were several contributing factors, including a rapid rise in population, with a consequent decrease in the per capita arable land, and a smouldering antagonism against the ruling Manchu that had to be sustained by the Chinese. But increasingly it was the aggressive interference by foreign powers that brought China to its knees during the course of the nineteenth century and beyond, until its resurrection by the Communist Party in 1949. Led by Great Britain, with its commanding naval force, but joined to the extent allowed by their means by other European nations and then, increasingly, by the United States and Japan, China was forced into the one unwinnable war after the other, each time resulting in harsh terms of surrender. Major milestones on this downward path were the Treaty of Nanking (1842, loss of Hong Kong); Treaty of Tientsin (1860, loss of Kowloon), Japan's annexation of the Ryukyu Islands (1874) and penetration into Korea (1885), which until then had been a Chinese protectorate; the Chefoo Convention (1876); the Treaty of Shimonoseki (1895, loss of Taiwan and of the protectorate over Korea); and the defeat of the Boxer Uprising (1900). These all involved war indemnities, foreigners' exemption from Chinese courts, the reopening of the opium trade (which the Chinese government had attempted to stop), and, notably, the promotion of missionary activity throughout China. This linking of the Christian church to foreign oppression was bound to leave its mark.

However, these events, which had a devastating effect on the Chinese economy, also had their effect on Chinese society. Firstly, in the century leading up to 1912, there were numerous internal revolts and uprisings, some regional, but a few encompassing large parts of China. Some were inspired by foreign ideas or beliefs, some were encouraged by the disrespect shown by foreign powers for the Chinese government, and some were simply a result of the declining standard of living. Not only did they directly weaken the authority of the central government and tax its already stretched resources, but the campaigns to defeat them gave rise to a number of powerful generals that pursued their own political objectives, which further reduced the national cohesion. Secondly, through the interaction with the Europeans in the treaty ports, a small middle class of Chinese emerged that adopted European methods and became familiar with European thought and political and philosophical concepts. This led to an increasing conflict between a progressive and revolutionary south and a traditional north, based on its conservative landholding gentry and with its anchor point in the imperial government in Beijing.

In 1912 the Manchu emperor abdicated, and China was proclaimed as a republic, albeit with the strongest of the generals, Yuan Shikai as president. He was in reality a conservative and a supporter of the gentry, and with ambitions to reinstate the monarchy with himself as emperor. However, he died before he was due to become emperor, and the years that followed, until 1927, were marked by infighting between a succession of generals and a collapse of the political power of the Beijing government. Several provinces declared themselves independent, and in the south, the revolutionary party—the People's Party—based in Guangzhou under the leadership of Sun Yat-sen, was steadily gaining in strength. After the death of Sun Yat-sen in 1925, the new leader, general Chiang Kai-shek, embarked on a campaign against the north which, after much fighting and many compromises, was successful in establishing a "united" China under the dictatorial leadership of Chiang Kai-shek. It was the nature of these compromises—essentially a conservative turn with persecution of socialists and communists—that resulted in the separation of Chinese society into two irreconcilable camps and the inevitability of a civil war *à outrance*. This was interrupted only by their common stand against the Japanese invaders, but once the Japanese were defeated by the US, the war resumed, and by the end of 1949 the whole of mainland China was under communist control. The defeated troops and their families fled to the Chinese island of Taiwan, where they were not particularly welcome, but where they have stayed, more or less under US protection.

In the seventy years that have gone by since the founding of the People's Republic of China (PRC), the country has undergone a remarkable, but also very turbulent rise, from a downtrodden backwater at the mercy of the West and with an economy in ruins to the second-largest global economy and a major participant in international politics. This despite being hampered by the legacy of its revolutionary birth and the hero-worship of its leader, and when compared to how Russia has managed a similar transition, the Chinese transformation has been outstanding. But it would be unreasonable to expect that the experience of the last two hundred years should now be forgotten; today's China is justifiably wary of the West's intentions, and is under no obligation to fit into a global system (euphemistically called "the rule of law") based on a political and economic ideology created by the West. China's internal development has been, and is today, one of unprecedented scale and dynamism, and requires a management system tailored to meet the challenges of this development. The management is a difficult balancing act that involves constant vigilance and resolute action whenever required; quite the opposite to the *laissez faire* so admired by the neoconservatives. Today, China is a major member of, and contributor to, the international community, and should be welcomed as such. It has not used any military force in attaining this position, and if its commercial practices are at odds with Western norms, this is a matter for negotiation, possibly within an existing framework, for example, the WTO. The present hysteria surrounding the rise of China is unfounded in fact, but the longer it is promoted, the more likely it is to become reality.

According to the view of evolution presented in this essay, with society as an information-processing system, the risk of instability is measured by the deviation of the operation of the collective intelligence from its optimal operating point. That operating point is characterised by a number of parameters, several of which were listed in Sect. 4.3, but the by far most volatile measure, and therefore the one determining the evolution in the short term, is the extent to which information is unrestrictedly available. The struggle to maintain the current non-equilibrium situation and the hegemonial position of the West is therefore, as we would expect, mirrored in the effort to ensure that the information available to not only the people on the Western world, but to as large a part of the world population as possible, presents a narrative that justifies that position. And not only justifies, but makes the western ideology, with its mixture of economics, ethics, and religion, as presented in Sect. 7.2.2, appear as the only viable one, as the final socioeconomic framework on which the further existence of humanity is to be based. We can discern three distinct components of this effort. The first one is a "filtering" or "editing" of information, in the sense of emphasizing, deleting, and associating items of information to create an impression in the reader's or viewer's mind that is favourable to the West and/or detrimental to what it considers its adversaries. For example, including the lifting one billion people out of poverty (<US$1.9 per day) in a list of achievements of capitalism, even though the majority of these were in China. Or suppressing the results of US-led airstrikes in Syria. Pictures of the devastation in Syria only started appearing very frequently in Australian newspapers once they could be assigned to Russian airstrikes. In the recent Houthi drone strikes on Saudi oil facilities the fact that the drones and rockets were manufactured in Iran was taken as proof that the attack was launched from Iran; whereas when a school or hospital in Yemen is hit by Saudi air strikes, it is not pointed out that the rockets are of US manufacture and that the strikes therefore are a US responsibility.

The second component is the extensive reporting of incidents in countries not aligned to the West that are contrary to western ideals of justice and freedom, even though they may be justified under the rules and customs of these countries. Conversely, when such actions occur in countries aligned with the West, such as Saudi Arabia or Egypt, there is little, if any, notice taken. In the case of China, two issues stand out in this regard, and the first is the role of religion and religious freedom. These are areas where the West sees an opportunity, cloaked as a moral obligation, to denounce any action on the part of the Chinese state that can be construed as violating the normative Western view of these areas. While useful in stoking anti-China feelings, this activity is unfounded and ignores the salient facts, which include the following:

1. Throughout its long history, organised religion in the form of a Church, has seldom been a significant part of Chinese society. Apart from early periods, when Buddhism was influential in some regions and Buddhist monasteries became very wealthy and powerful, the Chinese adopted a Confucian, pragmatic philosophy as a "state religion", and would, if anything, be justified as regarding Christianity and Islam to still be mired in superstition.
2. Through its role in the West's arrogant treatment of China in the nineteenth century, the Chinese could rightfully view Christianity as a Trojan horse for Western interference in China's internal affairs.
3. The history of the Christian Church is not one to inspire confidence in its benefits and behaviour. When it had the power to do so, it ran what we would today consider a terrorist operation; eagerly participating in the enslavement and extermination of indigenous people and amassing great riches in the process, banning the publication of any text not approved by it, and suppressing scientific knowledge and any opposition or contrary opinion by torture and burning at the stake, with the Spanish Inquisition alone burning more than 10,000 persons. And even today, the revelations of sexual misconduct by the clergy is shameful. So it is understandable and reasonable that China is sceptical as to the advisability of allowing the Christian Church to gain influence.

The second issue is that of freedom—both the freedom *to* and the freedom *from*. Regarding the first, any society places numerous restrictions on the freedom to act, and if we would measure these restrictions in terms of the number of laws, rules, regulations, taxation office determinations, and so on, it is not clear that the Chinese are any less free than, for example Australians. It is just that the restrictions are different, and it is in the framework of our restrictions that we judge some of those in China to be objectionable. The treatment of persons who advocate Western ideals in China is decried as violations of human rights, whereas people who promote the PRC in a country like Australia are seen as security threats. The difference between the two frameworks can be seen as a reflection of what we introduced, in Sect. 2.3, as a primary characterisation of society: how the members of society understand the relative importance of their dual roles as individuals and as elements of society as a system. This determines the extent to which they are willing to forego some aspects of personal freedom in exchange for the benefits arising out of increased integration.

Regarding the freedom *from*, we would probably first think of freedom from starvation, poverty, and ill health; then freedom from slavery, persecution, and oppression; and then, increasingly, from mental limitations, such as censorship. One concept—the standard of living—tries to combine many of these, but that inevitably involves both a selection of characteristics and their weighting; a simple selection is income per person and poverty rate. The Index of Economic Freedom is an annual index and ranking created by The Heritage Foundation and The Wall Street Journal, but it is firmly anchored in the neoconservative ideology, in that it assumes that the strength of basic institutions that protect the liberty of individuals to pursue their own economic interests result in greater prosperity for the larger society. The simplest measure is the per capita GDP, which is about US$60,000 for the US and about US$17,000 for China, with most Western European nations in the range US$50,000–US$60,000. A slightly more sophisticated measure is that proposed in [5] as a financial restriction, with 0 being best and 1 being worst; on this measure the US is rated 0.512, China 0.68, and Western European nations in the range from 0.418 (Switzerland) to 0.634 (Portugal). In any case, as far as the freedom from want is concerned, it is clear that the average person in the West is considerable better off than the average person in China; to what extent this can be attributed to their respective ideologies is a different matter, given our previous consideration of their histories. Of greater significance is that the rate of increase of any of these measures over the last twenty years or so has been significantly greater in China than in the West.

An illustration of the selective handling of information about China and the Chinese is provided by a comparison with the handling of information about Israel and the Jews. Both the Jews and the Chines were demonised as second-rate humans or "Untermenschen", and caricaturised as such in Western media. They were persecuted in the century leading up to WW2 and then subjected to unspeakable atrocities in the war—the Jews to the Holocaust, with its 8 million victims, and the Chinese to the Japanese terror with its 20 million victims. Both would have wowed that this would not be allowed to happen again, and took corresponding steps—Israel by acquiring nuclear weapons, encroaching on Palestinian land, and annexing the Golan Heights; China by modernising and expanding its defence capability, strengthening its grip on such border provinces as Tibet, and by annexing islands in the South China Sea. In the Australian media there is a report about some information item—interview, book, film, presentation, exhibition—relating to the persecution and discrimination of the Jews at least once a week, and there are also numerous items about Nazism and Hitler, and about antisemitism. But any similar items about the persecution and discrimination of the Chinese, or about Japanese imperialism and Hirohito, are practically non-existent, and the terms anti-sinoism or sinophobia are hardly ever used. The implication is that whatever happened to the Chinese was well deserved, and any criticism of the Chinese is not based on prejudice, but should be accepted as factual.

To get back to the question of international stability, the point is that the West and China are quite different—in their histories, in their current stages of development, in their ideologies—and they are competitors for power and for the greatest slice of the cake. But, being together on the same planet, they are also highly dependent on each other and, in the long run, must fit into a common framework together with all the other nations in the world; this is the next species of the genus society. The current polarisation, in the form of an intense propaganda effort to paint China as a dangerous enemy instead of simply another competitor, driven mainly by the West for the purpose of clinging to its hegemonial position, greatly increases the possibility that the path to this next species will involve a violent fluctuation.

References

1. von der Schulenburg M (2017) On building peace: rescuing the nation-state and saving the United Nations. Amsterdam University Press, Amsterdam
2. Chomsky N (2011) How the world works. Penguin Books Ltd., London
3. World Bank (2018) The World Bank annual report 2018. Washington, DC. https:// openknowledge.worldbank.org/handle/10986/30326. License: CC BY-NC-ND 3.0 IGO
4. Eberhard W (1960) A history of China, 2nd edn. Rutledge & Kegan Paul Ltd., London
5. Aslaksen EW (2018) The social bond: how the interaction between individuals drives the evolution of society. Springer, Berlin